Economics: a foundation course for the built environment

Also available from E & FN Spon

A Concise Introduction to Engineering Economics
P. Cassimatis
Paperback (0 419 15910 X), 384 pages

Construction Tendering and Estimating
J.I.W. Bentley
Paperback (0 419 14240 1), 250 pages

Property Valuation: The Five Methods
D. Scarrett
Paperback (0 419 13780 7), 232 pages

Construction Contracts: Law and Management
J. Murdoch and W. Hughes
Paperback (0 419 17060 X), 432 pages

Other titles

Arbitration Practice in Construction Contracts
D.A. Stephenson

Architectural Management
M.P. Nicholson

Building Economics: Theory and Practice
R.T. Ruegg and H.E. Marshall

Building Regulations Explained 1992 Revision
J. Stephenson

A Concise Introduction to Engineering Economics
P. Cassimatis

Construction Conflict Management and Resolution
P. Fenn and R. Gameson

Effective Speaking: Communicating in Speech
C.C.R. Turk

Effective Writing
Improving scientific, technical and business communication
C.C.R. Turk

Good Style
Writing for science and technology
J. Kirkman

The Idea of Building
Thought and action in the design and procurement of buildings
S. Groak

For more information about these and other titles, contact The Promotion Department, E & FN Spon, 2–6 Boundary Row, London SE1 8HN, Telephone 071 865 0066.

Economics: a foundation course for the built environment

J.E. Manser

Southampton Institute of Higher Education
Southampton, UK

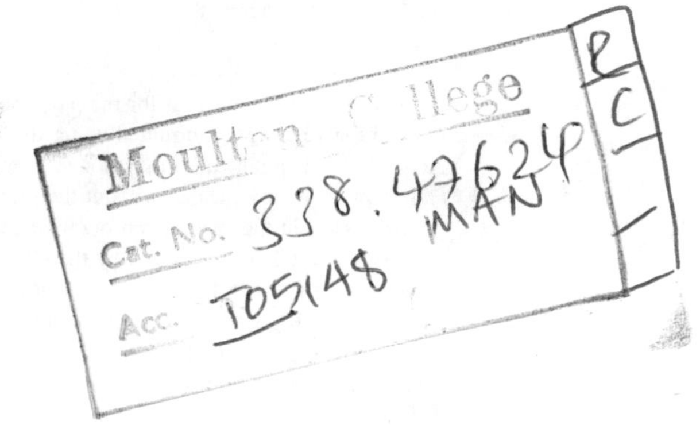

E & FN SPON
An Imprint of Chapman & Hall
London · Glasgow · Weinheim · New York · Tokyo · Melbourne · Madras

Published by E & FN Spon, an imprint of Chapman & Hall, 2-6 Boundary Row, London SE1 8HN, UK

Chapman & Hall, 2-6 Boundary Row, London SE1 8HN, UK

Blackie Academic & Professional, Wester Cleddens Road, Bishopbriggs, Glasgow G64 2NZ, UK

Chapman & Hall GmbH, Pappelallee 3, 69469 Weinheim, Germany

Chapman & Hall USA., One Penn Plaza, 41st Floor, New York, NY10119, USA

Chapman & Hall Japan, ITP-Japan, Kyowa Building, 3F, 2-2-1 Hirakawacho, Chiyoda-ku, Tokyo 102, Japan

Chapman & Hall Australia, Thomas Nelson Australia, 102 Dodds Street, South Melbourne, Victoria 3205, Australia

Chapman & Hall India, R. Seshadri, 32 Second Main Road, CIT East, Madras 600 035, India

First edition 1994

Typeset in 11/13pt Bembo by Gray Publishing, Tunbridge Wells, Kent
Printed and bound in Great Britain by Redwood Books, Trowbridge, Wiltshire

ISBN 0 419 18260 8

The use of the terms his and man is not indicative of specific gender and should be taken tomean his/her and man/woman throughout.

A Catalogue record for this book is available from the British Library

Contents

Preface

There are many introductory books on economics. There are fewer introductory books on construction industry economics. Construction students studying the subject often complain that it is not relevant, it is too abstract and does not help them in their working lives. In part this criticism is based on a misunderstanding of what is being attempted. Economics attempts to understand how things work, not to prescribe recipes for making them work better. But in part the complaint is well founded. The construction industry is in many ways unlike the generalized 'industry' which appears to form the model for standard introductions to economics and construction students therefore find it difficult to relate economics to their sector of the industrial world. My aim in writing this book has been to bridge that gap. I have tried to provide an account of the basic principles of economics and to indicate how these link to the real world of construction.

Initial chapters focus on economic theory; later chapters are more descriptive, setting the theory in the context of the industry. Throughout the book illustrative material is drawn from newspaper and journal articles reporting on the industry. These case studies are an important element in bringing together the two strands, theory and practice.

Although the book is specifically aimed at members of the industry who need an introduction to economics for professional examinations, I hope it may also interest readers who wish to extend their understanding of the economic context in which the industry operates. Having said firmly that this is not a book about business management, I nonetheless believe that a better understanding of the economic environment must benefit effective decision-making. The following suggestions on how to use the book are intended for students preparing for an exam in the subject, but other readers may like to adopt a similar approach.

To M.M. and K.R.

Using this book

PREVIEW

- How is the book laid out?
- What is the purpose of each section?
- How can I use it to study effectively?

You will find each chapter starts with half a dozen questions. These highlight the key issues which are going to be discussed. The questions are labelled 'preview' because you should look at them before you study the chapter; research has shown that people learn more easily when they have a clear idea of what they are looking for. The questions are to point you in the right direction and start you thinking about the issues. So spend five or ten minutes in preparation. You could take a piece of paper and jot down your immediate thoughts – just a few words will do – in answer to the questions. (Some you may not be able to answer at this stage.)

Ideally you should skim each chapter first to get an idea of the content, then read more thoroughly for a proper understanding. As you read you will see a number of terms are introduced in bold print. Many of these are defined in the Glossary at the back of the book, which is designed to give you a quick reference section to guide you through the terminology of economics. Jot down a few notes to highlight the main points. Taking notes is useful in a number of ways. It makes you think about what you are reading; it helps you organize your thoughts; it provides you with a summary for revision. Notes should be brief – headings, perhaps with one or two examples – and should not merely repeat sentences from the text. Expressing points in your own way ensures you understand them. Use numbers, underline or use a different colour to pick out the key points. Leave plenty of space so you can add ideas which you may come across later on.

With each chapter you will find case studies and exercises with questions to help you think about what you have learned. I do not suppose many people will have the time to answer all of these! All the same, looking at the questions will help you to focus on the way in which economics operate in the construction context. The case studies themselves are a vital element; most of them are drawn from the construction and property industries. Try to incorporate examples from these in your notes. (Better still, build up a file of your own examples, from cases reported in trade journals etc. or from your own experience working in the industry.) If you attempt some of the questions as well it will help you to practise the skills of organizing your thoughts and getting them down on paper, which are a necessary part of passing exams. Questions which are marked * are the most suitable for exam practice. Try to bring together the general principles in the body of the chapter with more specific instances drawn from the case studies. As you progress you may find you can draw upon earlier case studies to illustrate points in later chapters. The world does not come neatly packaged in individual chapters.

When you have gone through the text and the case studies, look back at the preview questions. How would you answer these now? Take a minute to jot down a few headings and compare them with the notes you made before you started. You will be surprised at your progress!

Finally I urge you not to regard this subject as something that should be confined to a textbook. We live daily in an economic world. The more you read the papers, listen to the news, or discuss issues related to your work, with an awareness of the economic implications of what is happening, the more relevant and enjoyable you will find your studies to be.

Acknowledgements

I would particularly like to thank colleagues in the Built Environment Division, past and present, for their cheerful support. I have enjoyed and benefited from working with Alan Johnson and I would like to thank him and Peter Gardner, both of whom took the time to read an earlier version of the book.

Most of all, my thanks to my husband whose fortitude and computer skills have been severely tested and not once found wanting.

The writer of any textbook owes a debt of gratitude to the many authors who have acted as pathfinders. The most important of these are listed in 'Further reading'. Since the case studies are a vital element of this book I am particularly grateful for permission to reprint them. I am also grateful for permission to use official statistics published by Her Majesty's Stationery Office (HMSO) and the Central Statistical Office (CSO).

1 Why economics?

PREVIEW

- How does economics involve me?
- What are 'resources'? (Is money a resource?)
- Can 'scarcity' be a problem if you are rich?
- What is the **real** cost of deciding to use resources for a particular purpose, such as building a house?
- How can we evaluate the production possibilities open to us, as businesses or as communities?

Economics is about making choices. Individuals, groups, businesses and nations are constantly having to choose how to use their resources. Can I afford a holiday abroad this year, or would it be better to get a new car? Should the firm invest in the latest in mini-excavators, or would it be better to buy more scaffolding? Is the Department of Transport's bid for new roads going to succeed at the expense of the Education Department's claim for more schools? If we had sufficient resources to satisfy all our wants, economics would vanish, but we never have enough resources for everything we want to do. So we economize, we make choices.

It is not simply a lack of money that limits what we can have. The government could easily print more money, but this would not help us. Money is only an IOU – a banknote is a written promise to pay the bearer a certain sum. It has value only in so far as we can redeem it, which means in so far as we can exchange it, use it to buy things, goods and services. It is our enjoyment of these which determines how well off we are. Standards of living are measured largely in terms of consumption.

What we consume depends not just on incomes, but on what is produced. If the government were to pass a law to double everyone's income there would not be any more goods in the shops for you to buy. You might have more money to spend, but so would everyone else. Stocks of goods would soon start to sell out and suppliers would raise their prices. In real terms we would be no better off.

SCARCITY, CHOICE AND COSTS

A higher standard of living means producing more goods. But production cannot be increased indefinitely. Limits are imposed by what **resources** are available, that is the skills of the workforce, the machinery they use, the raw materials, the energy and all the other inputs necessary to making the things we want to enjoy. (Sometimes resources are not fully used, in which case our economy will produce less than it could, but there is still a limit on what is possible.)

Limits on the resources available force us to make choices. More of one thing means less of another. This is what economists call **scarcity**, meaning simply that there is not enough to allow for everyone to consume freely as much as they would like. In cases where this sort of plenty does exist, for instance the air we breathe, no choices are necessary. Such items are called **free goods**. Free goods may be vital to our well-being, but they are not important to economics because production and consumption do not involve sacrificing alternatives. If I put my tomatoes in the sunshine to ripen the fruit, it does not mean there is any less sunshine for you to enjoy. Some free goods, like clean air, we take so much for granted that we are in danger of misusing them to the point where they may cease to exist in sufficient plenty for all to enjoy! At this point they too become scarce and their use poses economic questions.

Rainfall is another example of nature's bounty, which may be treated as a free good where it is in ample supply. The rain that falls freely on my crops also falls on my neighbour's – but a decision to build a storage tank for the rainwater means that it ceases to be a free good. It now becomes a stored water supply and so enters the realm of economics, of choices between alternatives.

Decisions have to be made. How much time should be given to building the tank? What materials should be used? Where should it be placed? How big should it be made? These questions are partly technical, involving issues such as the permeability of different materials and the design of the structure, and partly economic, involving questions about the most efficient use of resources. What is the cost and availability of different materials? How much time is it worth spending on the project? Time and materials used to build the tank cannot be used elsewhere. Once the task is completed, a free good, rain, is transformed into an **economic good**, something which is scarce, namely a stored water supply.

Further choices have to be made concerning the distribution of the goods. Who will benefit from this water supply? One basic decision about its allocation has already been taken: by storing the water, run-off is reduced, so there is more water for the household with storage tanks and less for households lower down the valley. Very often when we make choices that will benefit ourselves we find those choices also affect other people, sometimes adversely.

How shall the water be used? For irrigation, for drinking, for cooking, washing and cleaning? The supply is limited, the potential uses almost endless. Every litre put to one use means a litre less for some other use. Making choices implies making sacrifices, going without the alternative possibilities. This is an inevitable consequence of scarcity and economists use the term **opportunity cost** to emphasize this fact. Building the tank used resources – labour, materials and space – which could have been used in other ways (to build a grain store perhaps, or hen houses) but which are now committed and no longer available for other uses. Hence the water supply was created at the cost of sacrificing other possibilities: it has an opportunity cost; there is a price to be paid for the water – it is no longer free.

Students are generally very familiar with opportunity costs. Expensive textbooks compete with rent demands and the attractions of a night out on the town for the student's limited income. More money would solve some problems, but not all. There is another scarce resource, time! Managing time is as much an exercise in opportunity costs as managing a budget – the 'price' of a night on the town may be a rushed assignment and lower grades.

THE PROCESS OF PRODUCTION

Resources used to produce goods and services are called **factors of production**. They are classified under three headings: land, labour and capital. **Land** includes all natural resources – not just the earth's surface area but also the mineral deposits, forests, fisheries and so on provided by nature. **Labour** is the human effort employed in production, including managerial and administrative work as well as manual labour. **Capital** is the stock of man-made resources which have already been produced. It is the accumulation of tools, machinery, buildings, transport systems, etc., that we have manufactured to help us work more effectively and so increase our current output. As such, capital itself embodies past decisions on opportunity costs. The task of deciding how to combine these three elements, what goods to produce and which production methods to use, is termed **enterprise**. Some economists classify this as a fourth factor of production, others regard it as no more than a subdivision of labour.

The whole process of production is designed to provide **utility**, which means it is aimed at satisfying our wants. Utility cannot be measured objectively because people's needs and desires are individual. A beefsteak has no utility for a vegetarian, nor a cigarette for a non-smoker. But wherever a want is satisfied, whether it is a basic need or something entirely frivolous but fun, utility is created. This means that production involves more than the manufacture of goods.

All manner of services which contribute to consumer satisfaction are part of the production process. Consumers in Southampton or Edinburgh cannot enjoy their favourite TV programmes on sets still stacked in a Welsh warehouse. Utility is not created until the consumer's wants are satisfied, so the transport, insurance, advertising, stockholding and display of the goods in shop windows are all part of the production process. Similarly direct services, such as banking, hairdressing, education or entertainment, are all forms of production, helping to meet consumer needs and satisfy consumer wants.

Economic questions face us daily. As individuals we seek to maximize utility. Limited incomes mean choosing among the many goods and services available. Businesses, seeking to maximize profits, also have to decide between alternative courses of action – from the basic choices of what sort of goods to sell, through to detailed issues of design and production, all of which will have implications for the way they use resources.

A builder may decide to concentrate his efforts on house building, thereby sacrificing the opportunity to work on shops, offices, etc. This is only the first decision of many. What type of house should be built, which materials should be used, are traditional building methods preferred, or would pre-fabricated units be better, should the property be sold or let? Traditional bricks and mortar construction requires a different mix of materials and more labour than timber-frame or pre-cast concrete. Costs will vary, a fact which has implications for who will buy the houses, or, in other words, for the distribution of the goods.

Economics is the study of this process of using resources to satisfy wants. Are we managing to maximize utility? Are we using our resources in the best possible way? Could we, by making different choices, achieve a higher standard of living? What constraints limit us? Who should decide how we allocate resources between competing industries, and how to distribute goods amongst different members of the community? A brief consideration of these questions will raise many issues fundamental to economics. It also raises questions about the relationship between economics and politics. We will return to this in the next chapter.

TRANSFORMATION CURVES

We have already seen that the production of all goods and services involves an opportunity cost. If we want more of this we must take resources away from something else and have less of that. If a builder chooses to construct sheltered accommodation for the elderly, the land, labour and materials he uses will not be available to build homes for first-time buyers. The decision to put in a bid for this contract rather than that one, if successful, will mean the firm's resources are tied up and they will not be able to take on other work for the time being. This may mean sacrificing some potentially profitable opportunities. The entrepreneur must constantly weigh up the alternatives.

A **transformation curve** is a graphic representation of this opportunity cost. Along the two axes are measured quantities of the alternative goods which could be produced from a given quantity of resources. A joiner, for instance, can use his labour, tools and materials to produce doors or windows. The transformation curve plots all the different combinations of output which the joiner can achieve, showing the 'trade-off' between the two goods as resources are transferred from one to the other. This type of analysis can be applied to an individual producer or to a whole industry.

Let us assume that windows take twice as long as doors to make, but the joiner can switch from one to the other without any loss of efficiency. Figure 1.1 shows that if all the time is spent on doors, production is 10 units per week. If it is intended to make windows as well there will be less

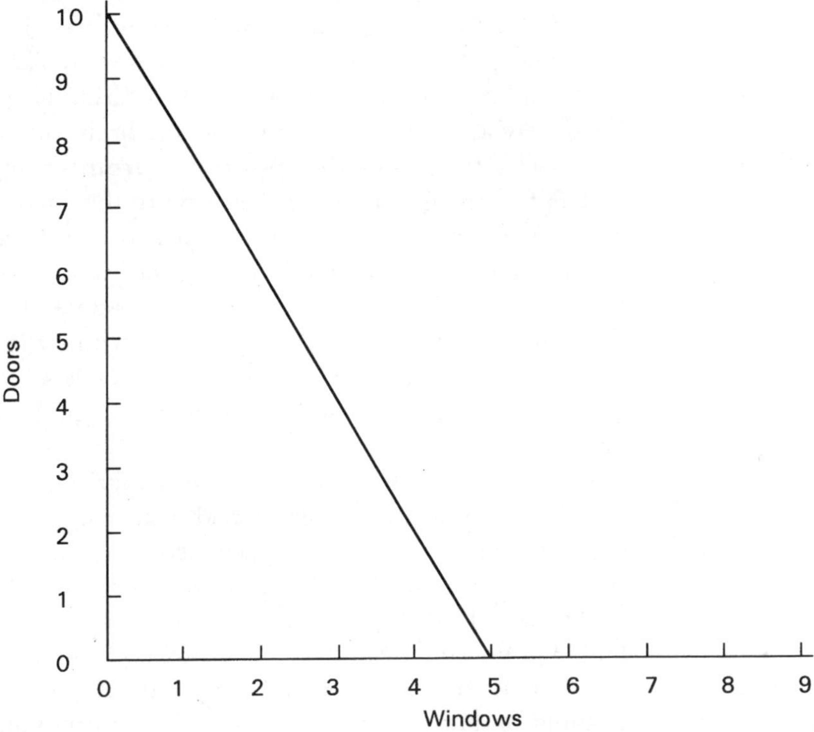

Figure 1.1
Transformation curve.

time to make doors and output of doors will fall. Producing one window means reducing the output of doors from 10 to eight, a second window will cut the number of doors to six, and so on.

This can be expressed as a transformation ratio of 1 : 2 (windows to doors) or 1 : 0.5 (doors to windows), which can be measured from the slope of the curve. (NB: a 'curve' refers to the line joining the points on the graph; it may be curved or, as in this case, a straight line.) Where the graph has a constant slope, i.e. it is a straight line, the transformation ratio will be constant, i.e. the same at any point along the graph. Whether current production is 10 doors or two doors, the firm will still have to reduce output by two doors a week if it wants to produce an extra window per week. If the firm wishes to produce more windows without reducing the output of doors it must obtain more resources, e.g. take on more labour, purchase additional tools, timber, etc. (This has knock-on effects elsewhere; more resources being used for doors and windows implies fewer resources available for other firms producing other goods.)

PRODUCTION POSSIBILITY FRONTIER

The analysis applied to a single firm in the example above can also be used to examine the options available to a whole economy. There is one important difference: a single firm can always increase its share of resources by out-bidding its competitors, but the resources available to the economy as a whole are more or less fixed in the short term. The 'transformation' curve for society as a whole thus represents the limits on production and is more often called the **production possibility frontier** (Fig. 1.2).

The first thing to notice is the shape of the curve: it is no longer drawn as a straight line. Why is this? The firm that was making doors and windows could make either product using the same labour, the same materials, the same machinery, the same skills. We assumed that the resources could be transferred from one product to other without loss of efficiency. When we look at broader categories this becomes less realistic. In Fig. 1.2 we are looking at the trade-off between more agricultural products or more building work.

Resources, whether land, labour or capital, are not uniform in quality. Some may be highly specialized, and if they are switched to a different use there will be some loss of productivity. Agriculture and building differ in the sort of land, skills and capital equipment which they use. If we start with most of our resources devoted to agriculture and very few given to building, we could construct more buildings for very little loss of agricultural output at first. We would start by developing the sites that were easy to build on but not the most fertile, so as to minimize loss of agricultural

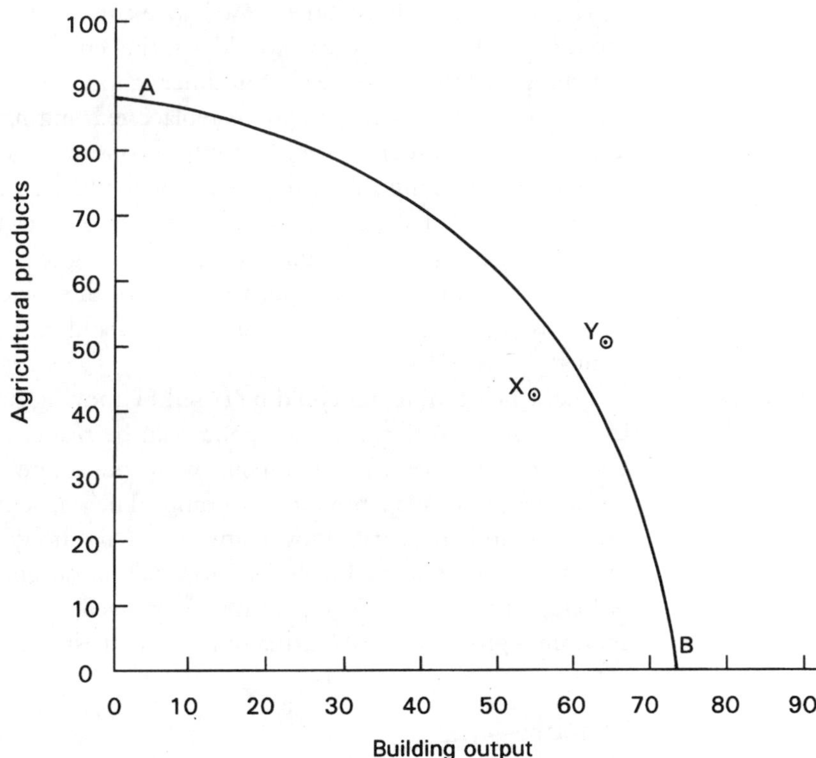

Figure 1.2 Production possibility frontier.

output. Similarly we would use the best builders, not the best farmers, so as to get the most construction output. But as we move along the curve, away from A towards B, more and more land and workers are being transferred to building, until only the richest farmland and the labour least willing to leave agriculture remain. If they too are transferred the gain to building will be small, since they are not well suited to the industry, and the loss to agriculture will be great. Hence the sacrifice of agricultural output for building output changes as we move along the curve.

Once again the slope of the curve gives a measure of this transformation ratio between A and B. (To measure the slope at any point is a simple matter of drawing a tangent at that point and taking the slope of the tangent.) Any point along the curve is a possible point of production, given the resources available. If we are not on the curve, but at some position inside it, e.g. 'X', then we are not using the resources of our economy to the full. It would be possible to increase production of either A or B without any sacrifice of the other, or even to increase output of both A and B. Any point beyond the curve, e.g. 'Y', is beyond the frontier of possibility, given the available resources.

This type of analysis can be used to examine the opportunity costs of production decisions. What would be the effects on living standards of switching resources between consumer goods and capital goods? Some capital goods are needed simply to replace existing plant and equipment as it wears out, in order to keep output of consumer goods at present levels. Suppose maintaining the current level of production requires an output of 50 units per capital goods, so that the rest of society's resources can be devoted to producing 57 units of consumer goods (Fig. 1.3). If all the resources available are being put to use, then the society, at point V on the curve, is enjoying the highest, sustainable standard of living which it can achieve.

The standard of living could be raised by moving further up the curve to U, but this means fewer capital goods can be made. As factory production lines and other capital equipment wear out there will not be enough replacements to keep production going. The capacity of the economy is shrinking and the curve moves inwards. This shows that too much consumption now will eventually lead to a falling output and **lower** standards of living in the future. The gain was short lived. If society wants long-term economic growth, it must actually increase its stock of capital goods. This

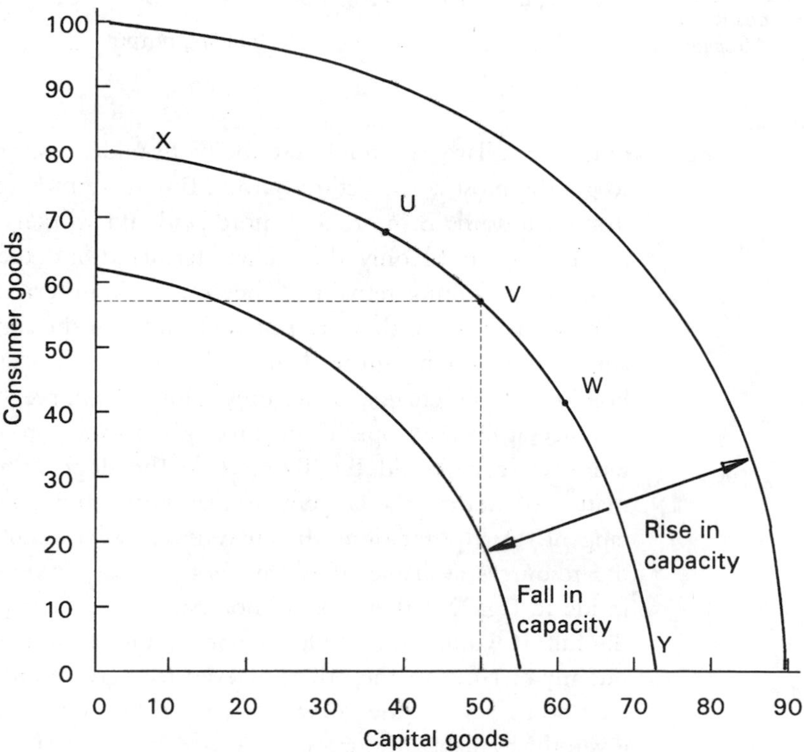

Figure 1.3
Shifts in production possibility frontier.

may mean moving down the curve, e.g. to point W, to a lower level of consumption at present. In other words, we must first **save**, that is we must release resources by not using them for consumer goods, then we must **invest**, that is use those resources to increase our stocks of capital goods. In this way the frontier will gradually be pushed outwards as the economy's capacity to produce grows bigger.

A NOTE ON TERMINOLOGY

Some readers may find familiar words being used in unfamiliar ways when they read economics. The above passage refers to saving, investment and capital, all everyday words which have become part of the specialist language of economics.

What does saving mean to you? Probably putting some money aside instead of spending it. This is very similar to the definition used above – 'not using them (resources) for consumer goods' – except that the latter was used in the context of real resources (labour, materials, plant) rather than financial resources (money). In both cases it implies a cutting back on present consumption. Similarly investment in everyday terms refers to what you do with your money, to bring in an income or increase your wealth. Investment to the economist means using the resources saved to create new assets (capital goods) that will help increase our output, and create long-term economic growth.

Many of the particular terms of economics are ordinary words given more specific meanings. Scarcity was another such term, encountered earlier. In everyday language this implies a particular shortage; in economics it means simply that demand would exceed supply if the price were set at zero. Look out for these as you study, and in your reading about economic affairs.

It is important to define terms as carefully and exactly as possible in order to make clear distinctions and logical arguments. Look for the connections with everyday language as well as the differences. If you can explain the terminology of economics clearly, you will have grasped the concepts and be well on the way to following, and then criticizing, the arguments!

CASE STUDY

A. Guns or butter?

Whilst the Bible enjoins us to beat swords into plough shares the politicians are in turmoil over the consequences. As the former Soviet empire disintegrates Whitehall is wondering how the UK should respond, in particular what to do with the £23 bn earmarked for defence expenditure.

Over at the Treasury officials are keen to keep government spending in check and welcome the prospect of a £1 bn 'peace dividend'. This would help to keep local taxes down and give a boost to health and education spending. They gained unexpected support from the Minister for Defence Procurement who declared in a radio interview that 'very substantial' cuts in defence spending were being planned.

This could mean as many as 23 infantry battalions being disbanded, with a loss of 50 000 men for the Army. The Navy might lose a third of its frigates and destroyers, plus the likelihood of the fourth Trident submarine being cancelled. Up to half of the civilians at the Ministry of Defence could be in line for redundancy, while new anti-tank radar systems and the Challenger II battle tank are also under threat.

WORKSHOP

For questions 1–3 refer to case study **A**.

1 What is meant by the 'peace dividend'? What economic concept is implied by this phrase?

2 Consider the economic implications of cuts in defence expenditure of the sort suggested above.

3 In the light of your answer to 2 would you expect £1 bn less in defence spending to produce an equal sized benefit in health and education, assuming the government used the savings for these services?

4 A developer has acquired a 5 hectare site for housing. Building detached houses provides 40 dwellings per hectare; blocks of 4-storey flats give a density of 96 dwellings per hectare.

a Draw a transformation curve to show the combinations of flats and houses which could be achieved on the site.
What does the shape of the curve indicate?
If the developer decided to build 120 houses, how many flats could be built in addition?

b What factors might the developer take into account when deciding on the ratio of houses to flats?

DISCUSSION QUESTION

❑ How are the questions posed by economists about the use of resources reflected in the problems faced by builders and developers in their businesses?

2 Economic systems

PREVIEW

- What are the main problems facing a society as a result of scarcity?
- If you are free to buy what you choose, will this mean you can get what you need?
- Can a central authority with powers to direct resources resolve the problems of scarcity?
- If there is no central authority, how do we avoid producing too much of some goods and not enough of others?
- Who should decide how a society's resources are used: producers, consumers, economists or politicians?

The fundamental questions that any society has to answer, concerning the use of its resources, are:

- ***What** goods and services shall we produce?*

  Do we want more roads, more cars, more houses, more hospitals, more videos, more ... of everything? We know we cannot have everything, so we must put our wants in order of priority and accept that more houses, for instance, will mean less of something else.

- ***How** shall we produce them, which resources should we use?*

  Shall we use traditional, labour-intensive methods (and employ more people) or shall we opt for modern, automated methods (and produce more goods)? Should we develop land intensively, building high-rise blocks, or do we prefer low-density developments, even though it uses more land and means more built-up areas and less countryside?

- ***Who** is to have the goods and services which we produce?*

  How are goods to be distributed? Are they to be sold to the highest bidder, according to willingness to pay? Or should they go to the neediest, or simply to the first in the queue, or be shared out equally? There are many criteria and we may use more than one method to distribute goods, according to the circumstances. Housing is allocated according to ability to pay in the private sector, with an element of 'first come, first served' when shortages occur. The public sector combines criteria of need with a queueing system.

Different societies have approached these basic economic issues in different ways. Traditional societies, of which there are few left in the modern world, relied on the patterns of work and distribution established by their forefathers. In a pre-industrial society, where subsistence depends largely on the cycles of the natural world, the experience of generations is a sound basis for decisions. Production depends on the relatively unchanging rhythms of nature, when to sow and when to harvest, where to hunt, how to build with the materials to hand. Distribution is determined by traditional social structures. The community generally has little surplus after its basic needs have been met; distribution is more likely to be a problem arising from crop failures and shortages than a question of sharing wealth.

In the modern world of innovation and technology, there is a larger range of choices to be made and past experience is no longer an adequate guide. Paradoxically, greater wealth may also create more problems concerning the ways it should be shared out. Economic decisions are inevitably bound up with political and social decisions. A 'free' society, one which values the rights of individuals, must allow freedom of thought, of political and religious views; it must equally allow freedom of economic choice.

MARKET ECONOMIES

Economic decision-making in such a society will be decentralized. **What** goods are produced is decided by individuals who choose how to use their labour and how to spend their incomes. Most resources, i.e. land, labour and capital, are privately owned and the owners are free to sell to the highest bidder. **How** resources are used depends on the prices offered for them by competing producers. **For whom** depends on consumers' willingness and ability to buy the goods on offer.

The consumer looks at the goods in the market-place and decides what to buy. Millions of individuals make millions of choices, each trying to maximize personal benefit (utility) from their choices. How can we be sure that this will result in the 'right' amount of bread and potatoes, caviar and champagne being produced? Or the 'right' amount of bricks, blocks and bulldozers? How can producers judge what consumers will want? It is the 'invisible hand' of market forces, acting through the price mechanism, which informs and coordinates individual choices. How this works will be examined in more detail in the following chapters.

This sort of **market economy**, or **free enterprise economy**, has much to commend it. It allows us to choose what we want. Variety and innovation are encouraged. As consumers we find our wants are met because we are willing to pay the producers who satisfy us, and so producers, motivated by the desire for profit, seek to match their goods to our demands, to find more effective ways of pleasing us.

Those who succeed will survive and prosper, just so long as they continue to judge our wants correctly. Those who get it wrong will go out of business. The resources they controlled are then freed for the more successful firms to take over and expand their output. Theoretically at least, the customer is king – **consumer sovereignty** is the name of the game!

MARKET FAILURE

In practice there are many problems and markets sometimes fail to meet our needs. **Market failure** arises in a number of ways. There is the tension between freedom and equality, both of which are necessary to an efficient market. In economic terms freedom of enterprise will tend to produce inequality of wealth, because some people will be more talented, innovative, work harder, etc., than others. They will make profits where others fail. This means that individuals' purchasing power, and hence the share of resources which they can command, will also be unequal.

As a result some people will have incomes too low to afford an adequate amount of even the more basic needs, such as housing. An economic system which satisfies the whims of the seriously rich, whilst allowing the poor to drift into dereliction, is contributing to social divisions which may ultimately undermine that society. Luxury penthouses alongside cardboard shelters can create a sense of deprivation and destroy the respect for property which underpins a free enterprise system. Economic issues of distribution rapidly become political issues of justice. Economic freedom is a mockery for those too poor to have real choices.

Markets can also fail to give due weight to the needs of the society, as distinct from the individuals who make up that society. Some communal needs, such as defence, cannot be supplied by a market. If the nation's defence system keeps the peace, everyone shares equally in the benefit. I cannot choose to buy more or less peace than anyone else, nor to go without. Since few people would willingly buy something which they will enjoy whether they pay for it or not, such **public goods** must be provided collectively or not at all.

In other cases it may be possible to provide goods on an individual basis, but the benefits spill over into the community at large. A good education is something which many individuals are prepared to pay for because they recognize the advantages that it offers them. Others may be unable to afford it, or prefer to spend their money elsewhere. These decisions affect the whole society; if people are not well educated they will be less efficient, our industries will be less productive, and our standard of living as a nation will suffer. Education is a **merit good**, one which has social as well as individual benefits. Left to make our own choices, we may undervalue it and choose to consume too little.

PLANNED ECONOMIES

It seems the market does not automatically achieve the best solution to our basic economic questions. An alternative is a planned economy where decision-making is centralized. Resources are communally owned and a planning authority organizes production to meet the society's needs. This means collecting information about what sort of resources are currently available and how much there is of each type, deciding on the priorities amongst competing consumer needs, setting production targets for the various industries, allocating the appropriate resources to those industries and, finally, deciding how to distribute the goods amongst consumers.

Planning in the old Soviet Union showed the difficulties. Long-term objectives were set out in strategic five-year plans; short-term targets were fixed annually in plans detailing monthly production figures on an industry-by-industry basis. This complex process required millions of work hours to calculate. If projected production did not match predicted consumption the planners had either to alter output targets, or change consumption by altering people's purchasing power. Both prices and income were set by the planners. Production targets had to be matched to resources, labour trained and directed as needed. State banks were told how much finance to provide.

The most difficult task was to equalize the supply of inputs of raw materials, manufactured items and component parts, themselves the output of other industries, to meet the final output targets. The problems are illustrated in a simplified way with hypothetical numbers in Table 2.1.

Oil can be obtained from current production, from stocks or imported. (In the late 1980s production was around 600 m tonnes but by 1993 after the collapse of the Soviet system it had fallen to an estimated 350 m tonnes.) Oil is used as an input in petrochemicals, to generate power, to fuel transport, and by the oil industry itself. It is also sold to consumers for domestic heating and transport, and it is exported. Suppose the industry fails to meet its target. The planners can choose to use more stocks (if they exist), to import (which requires foreign currency), to cut exports (and lose

Table 2.1 Matching output and requirements (m tonnes)

Oil supply		**Oil required**	
Output from:		Intermediate consumption:	
Production	610	oil industry	5
Stocks	3	chemicals	25
Imports	10	electricity	38
		transport	420
		Final consumption:	
		domestic users	45
		exports	90
Total	623	Total	623

foreign currency earnings) or to cut the supplies to other industries (with implications for their contributions to the plan) or to domestic consumers. No wonder Soviet citizens became accustomed to shortages!

The strengths and weaknesses of planned and market economies are, in many respects, mirror images of each other. Markets are subject to inequalities, to cyclical recessions and booms and often fail to meet social needs. On the other hand they are dynamic, flexible and reward enterprise and efficiency. A planned economy can (though it does not always) ensure wealth is more equitably distributed, can take a longer-term view, and can maintain jobs for all, even if many jobs are not very productive. However, it also suffers from excessive bureaucracy, and is slow, cumbersome and lacking in incentives. Consequently it may be inefficient and offer a lower standard of living as well as imposing restricted choices on its citizens.

CHANGING CIRCUMSTANCES

In Eastern Europe we have seen one planned economy after another introducing market elements into their systems and rejecting the apparatus of centralized state controls. The great Marxist vision of a new social and economic order, based on the need of communities rather than the greed of individuals, has collapsed. It would seem that the defects of planning, as we have experienced it so far, are greater than the advantages. This should not lead us to reject planning outright, but to consider more carefully what it can and cannot do.

Where a society has well-defined and generally agreed objectives, central direction of resources may be the most efficient way of achieving those objectives. It is difficult to imagine a more thoroughly controlled economy than Britain in the Second World War. Labour was conscripted to work in factories, down the mines and on the land. Ration books were issued to consumers, entitling them to purchase limited amounts of food, clothing and other essentials.

The system for organizing building activity had to take account of the needs for airfield construction, essential factory space, repairs following bombing raids, etc. It was set up by the Ministry of Works and Building in 1941 and the House of Commons was told how planners had to estimate the total value of building work which could be undertaken within a given time period. They divided this between the various government departments, so that each had a share. In making the allocations they had to take account of how much labour and materials would be available and also set strict priorities, such as giving preference to constructing airfields over building houses.

Strange as that seems today, it is equally hard to believe that the war effort could have been effectively managed by any other means. To transform production from civilian to military goods on such a large scale, so quickly (many would argue it was not done quickly enough), and for such a prolonged period, required strong, central direction. This was acceptable because society was overwhelmingly agreed on the objectives of the war. In normal circumstances such a degree of centralized control would be unacceptable, because individuals differ in the ways they wish to use their resources.

So far planned economies and market economies have been described in terms of opposites, but the alert reader will realize that no economy can exist without some element of planning, nor can any economy ever be totally planned. In practice some mixture of market forces and central direction will be employed. Britain in the post-war era has often been described as a **mixed economy**. After wartime controls were removed we retained a considerable measure of centralization. Many resources were controlled by state-owned industries and the provisions of the welfare state helped to redistribute incomes more evenly.

In the 1980s Britain, along with other industrialized economies, moved further towards the market end of the spectrum. Major state industries have been privatized, markets have been deregulated (this does not mean the removal of all regulation, but greater opportunities for competition) and citizens encouraged to look for alternatives to state-provided services and benefits. The Victorian virtues of thrift, self-reliance, hard work and enterprise have been much lauded. Such qualities built the thriving economy of the nineteenth century but we must remember Victorian society also encompassed much poverty, degradation and misery. The balance between individual freedom and community provision needs constant review and must be adjusted to meet changing circumstances.

POSTSCRIPT

This section has dealt with broad classifications of economic systems. There is considerable overlap between economic and political systems, demonstrated in the analogy between free enterprise economies and the democratic states in which they flourish. There is an implicit political stance in the closing remarks, for which I make no apology. These remarks are **normative statements** – they deal with opinions as to what ought to be the case. Many economists, who wish to stress that economics is a science, albeit a social science, deny that normative statements have any place in the subject. Questions as to what ought to be are seen as the preserve of politicians and voters. Economists are concerned only with **positive statements**, statements which are factual, which are capable of being proved true or false.

The distinction between normative and positive is an important one. Economics is an objective study. It uses scientific methods: economists first look at how the world works and collect data about it. The next step is to formulate hypotheses that will explain these facts. Finally the hypotheses need to be verified. It is at the latter stage that economics diverges from the 'hard' sciences because it cannot set up controlled experiments to test its theories. We can only predict outcomes and wait to see if events unfold as we expect.

The complexity of the economic world leads us to build models, simplified conceptual systems by means of which we can examine variables one at a time. This often strikes students as unrealistic, but it is analogous to the modelling of particle behaviour by mathematical equations in physics. By using computer models to link large numbers of economic variables we can manipulate individual items and test the outcome in a manner similar to the hard sciences' experimental techniques. Unfortunately even computer models are limited in scope compared to a real economy and simulations of this sort are not entirely reliable. All sciences aim to achieve a better understanding of how the world works. Economics, like other sciences, seeks to establish chains of cause and effect. If soundly based, this knowledge will help us to predict the consequences of our decisions and thus to manage our resources more effectively.

When arguments which are based on facts and logical chains of causation become mixed with opinions based on value judgements, often unstated, much confusion results. Personal prejudices may be presented as proven laws of science, or economic facts. Hence the emphasis many economists lay on positive economics, i.e. verifiable data not opinions. Yet, if the subject is to be more than an intellectual exercise, fascinating but sterile, it must be because it helps us to make better informed decisions as consumers, as producers, as citizens. We need not abandon opinions but we must avoid confusing them with proven facts. As understanding deepens, we may then find our opinions changing.

CASE STUDIES

A. The economic ballot box

The free enterprise economy is the true counterpart of democracy: it is the only system which gives everyone a say. Everyone who goes into a shop and chooses one article rather than another is casting a vote in the economic ballot box: with thousands or millions of others that choice is signalled to production and investment and helps to mould the world just a tiny fraction nearer to people's desire. In this great and continuous general election of the free economy, nobody, not even the poorest, is disenfranchised: we are all voting all the time. Socialism is designed on the opposite pattern: it is designed to prevent people getting their own way, otherwise there would be no point in it.

Reproduced from Enoch Powell, Freedom and Reality; *published by Batsford, 1969.*

B. A moment in time

Each of the following items is based on reports appearing in a single issue of the *Financial Times*, published on 3 July 1990, the day after East and West Germany implemented their currency union.

1. Prospects of German unity

On the first shopping day after the D-Mark became legal currency, Wittenbergers queued for a multinational selection of fruit and vegetables and filed in front of shops arrayed with brightly coloured soap powders, toothpaste, toys, whisky and coffee ...

The homegrown entrepreneur was on show on the square in the form of Ms Gertrud Olschewski, a former Wittenberg waitress who has leased a mobile pizzeria stand from a company in Bielefeld, West Germany. By lunchtime she reported that sales had reached about 1,000 of the new delicacies at DM3.50 (£1.20) each. 'All the ingredients – cheese, ham, salami, tomatoes and the dough – are from the West,' she said. 'That's where all the good things come from.' ...

Another optimist was Mr Harald Berndt, the 33-year-old deputy administration manager of Potsdam's 18th century rococo Sanssouci palace. 'It's a great chance – I am ready to work harder,' he said, although adding that his DM850 net monthly pay would eventually have to go up in line with the salaries of custodians of other Prussian palaces ...

Pressures of a much more brutal kind are faced by Mr Wolfgang Sonntag, sales and export director of a state-owned chemical plant at Reinsdorf, a village on the road to Wittenberg. The company is preparing to lay off 250 of its 800 employees. The company has a good stock of orders ... but productivity is only two-thirds that of Western competitors and it has received no real capital investment for twenty years ...

Mr Ralf-Rudiger Hoffmann, a former Staatsbank employee [now managing the newly opened Deutsche Bank] underlined the mixed emotions. 'Satisfaction over the end of the old system was there already,' he says. 'But as the contours of the new system become more clear, fears are growing.'

2. Collapse of UK building companies predicted

Up to 30,000 small and medium sized building companies may go out of business this year as the recession in the construction industry deepens, Barclays Bank warned yesterday.

Mr Richard Roberts, the bank's construction economist, said bad debts in the industry could rise by 40% this year, increasing pressure on contractors and material suppliers. He said falling house sales and prices could cut net revenues of housebuilders by a fifth this year to the lowest level since 1982.

The construction industry traditionally has a high turnover of company formations and failures. During the 1980s about 20,000 construction businesses a year ceased to trade ...

[Mr Roberts] blamed high interest rates for the falling house sales and prices but said these had probably touched bottom. Even so the outlook for housebuilders remained bleak. Profitability could fall further next year even if sales and house prices improved on the back of lower interest rates. This was because builders still had to work through some of the expensive land bought during 1987/8 on the back of the then rising house prices ...

Developers and contractors to fail this year included Rush & Tompkins, Declan Kelly, Brims Holdings, J.M. Jones, Federated Housing and Stanley Miller. These failures have had a knock-on effect on small subcontractors and building material suppliers some of which have already followed these companies into receivership.

3. Entrepreneurship in the East Bloc

As east European countries grapple with the transition from planned to market economies, it is the struggle against officialdom which preoccupies Ladislav Vostarek, a Prague lawyer who is also a rock music impressario, a song-writer and head of the six-month-old Association of Czech Entrepreneurs ...

'We are suffering from a weight of bureaucracy,' he says. 'Bureaucracy is an obstacle between the people and the government ... Under their slogan of "We cannot allow anarchy in our economy," every activity must be checked and controlled. You need 13 official stamps if you want to form a company with permission to export, 10 if you try to start a simple business like a travel agency. You have to spend a lot of time queueing outside the doors of clerks in their offices. There is a 30 day wait for permits. You have to wait weeks to be entered on the official register. All safety and fire precautions have to be officially inspected ... We don't have a properly operating banking system to obtain loans. We live in doubt about whether raw materials will be available, and about how much tax we will pay at the end of the year ...

Florin Burada is the head of a new consultancy for small businesses in Romania. So far the government has issued 21,750 permits for people to start up small firms but only 11,270 have got going ... 'There is a problem because these small firms have asked for premises,' Burada says. 'Local town halls have not been dynamic in looking for suitable places for them ...

Poland has just formed a Bank of Social and Economic Initiatives. Pawel Wyczanski says 'We need trained people from enterprise agencies. We need advice from people who have done it. We are having to learn how to assess risk among "unbankable" people. The bank's main business will be to give extended credits to new businesses as well as to enterprises which can employ more people ...,

Jan Bielecki, newly elected to Szczecin's city council: 'People are disappointed ... How do we convince them we are right to foster entrepreneurship, training and small businesses instead of giving everyone pay rises? People felt safer in old-style, big state companies run in the old way ...'

Pal Reti, of the *World Economy Weekly*, which is published in Budapest ... sees a small irony. Small business development in the West has been encouraged and aided by governments, usually working in some sort of partnership with the private sector and local authorities. 'But in Eastern Europe there is a continuing withdrawal of the state from all sectors of the economy. The irony is that we need state help.' ...

WORKSHOP

For question 1, refer to case study **A**.

1 Is the passage a statement of positive economics or a normative statement?
Compose a short sentence, based on the passage, which makes a positive (factual) statement. How, in principle, could you set about proving/disproving the facts you have stated?
Write a short sentence expressing a normative view presented in the passage. How would you refute such an opinion?

2 Referring to Table 2.1, in place of the table using oil as an example, formulate a table balancing the requirements for steel with the sources of supply. How many intermediate industries require steel as an input? (Restrict your table to 3 or 4 main users). How much steel will be sold to final consumers?
If steel output falls 15% below its target, what adjustments would you recommend to the plans? What knock-on effects do you envisage? How would construction output be affected?

3 Read extracts 1, 2 and 3 in case study **B**. What light do they throw on the workings of **a** market-based, and **b** planning-based economic systems?

DISCUSSION QUESTION

❑ 'There is no place in a market economy for public sector construction organizations; efficiency can only be achieved by ensuring there is competition among private firms.'

3 The market mechanism

PREVIEW

- How do consumers decide which goods to buy?
- How do suppliers know what consumers want?
- How does a market manage to supply the right quantity of goods to meet the demand of buyers?
- What is the 'right' price for a product?
- What happens when prices change?

THE MARKET

Markets bring together buyers and sellers of goods. There need not be a face-to-face meeting, although many markets are conducted in this way. Equally effective markets are operated by currency dealers who buy and sell around the world without leaving their desks. The market is the trading, the negotiation between would-be buyers and sellers, that takes place in it. It is most obvious in the face-to-face haggling of a street bazaar, or its modern equivalent, a car boot sale, but deals made via a computer screen are essentially the same process.

Prices are the market's communication system. Changes in price signal shortages and gluts, giving traders information about the availability of the goods and the extent of the demand for them. Decisions to buy or sell depend on price. Prices determine how resources are allocated between firms, and goods distributed amongst consumers. The price mechanism is the 'invisible hand' that guides market decisions. To see how this works we will use a model called the **perfect market**. ('Perfect' does not mean it is the best market. It means we stipulate certain conditions to prevent accidental factors hiding the essence of the market's workings.)

These conditions are:

1 the existence of **many buyers and sellers**, so that no individual person or firm has enough influence to fix the price;
2 the product traded is **homogeneous**, i.e. of a uniform type and quality, so there are no preferences based on variations in what is offered for sale;

3 everyone has a **perfect knowledge** of any market changes and can respond quickly to them;
4 there is **no external intervention** by outside bodies to impede the free operation of market forces;
5 there is **freedom of entry and exit**, so that anyone can buy or sell in the market and there is no risk of cartels or price rings excluding competition.

In addition there are the assumptions common to most economic theory, of **rational behaviour** (that people pursue their own best interests in a consistent manner) and of ***ceteris paribus*** (that everything, apart from the variable under scrutiny, remains constant). Although this model is rather artificial it allows us to focus on the most important aspect of a market – how prices are determined.

DEMAND

Effective demand means more than just desire, or even need, for a product. The homeless need shelter, the tenant dreams of home ownership, but this is not enough to create a demand unless they have the money to make good their wishes. Since consumers have only a limited amount of money available, it is clear that the price of the goods will be important in deciding whether to buy and how much to buy.

If the price falls some consumers will buy more, and others who could not afford the product will now be able to. If the price rises there will be fewer purchases as some buyers cut back and others cease to buy. Thus there is an inverse relationship between price and quantity demanded. Unless we know the price we cannot quantify the demand. (Naturally we must also specify a time period if we want to know how many units will be required by buyers.)

Quantity demanded can then be tabulated in a schedule of figures (Table 3.1) or shown graphically as a demand curve (Fig. 3.1). The curve shows the effects of a change in price, *ceteris paribus* (i.e. assuming no other changes), on consumers' willingness to buy. The effect of any change in price can be followed by moving up or down the curve to the new price level and reading off the quantity now desired. (NB: Prices are always plotted on the vertical axis.) Because demand varies according to the price level, we say demand is a function of price:

$$D f P$$

The graph shows that as the price rises fewer bricks will be purchased. If we want empirical evidence about the shape of the demand curve we can

Table 3.1
Hypothetical demand schedule: weekly demand for standard bricks

Price (£ per 1000)	Quantity ('000s)
230	100
220	200
210	300
200	400
190	500
180	600
170	700
160	800
150	900

look at sales records or conduct a market research questionnaire, but it is difficult in practice to isolate the effects of price changes from other factors which influence demand. So far our model has only looked at price changes.

Buyers may alter their plans for many reasons apart from price. A change in income will affect their ability to buy. Changes in tastes or fashion motivate some purchases. One of the most important influences on the demand for a product is the price of **other** goods, be they substitutes or

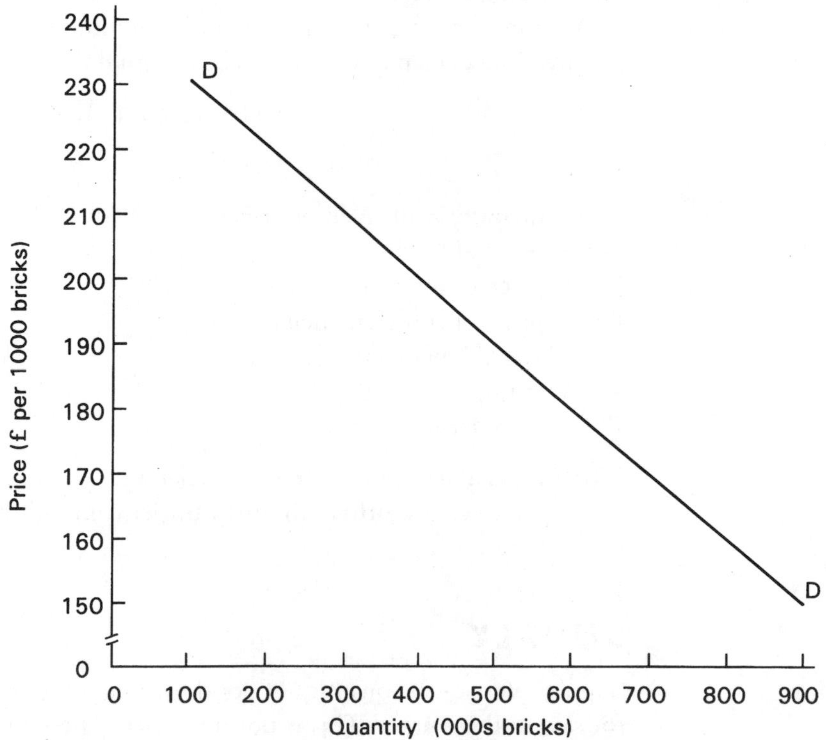

Figure 3.1
Demand curve (bricks).

complements. A **substitute** is an alternative to the product – the choice of meat or fish for dinner, or the choice between making a journey by bus or taxi. In the building context, slate or tiles are possible substitutes for roofing. (Can you suggest a substitute for bricks? Possible alternatives may vary depending on how the bricks are being used, whether for load bearing walls, for cladding, as a decorative finish, etc.) Decisions about which material to use are frequently made by comparing the prices of the alternatives. **Complementary goods** are those used together (films and cameras, CD players and CD recordings, bricks and ...?) A change in the price of one alters the costs of the 'package' and affects demand for both. If timber frame construction replaces traditional building and fewer bricks are used, the demand for mortar is also affected.

Here the change in the demand for bricks is independent of changes in brick prices. If there is a shift in demand which is not due to price, we say the **conditions of demand** have altered. To show this on the graph we must move the position of the demand curve. If demand increases, perhaps due to a housing boom, the curve shifts to the right, so that the existing price level correlates with a larger quantity; if demand decreases, the curve shifts to the left. (NB: Quantities are always plotted along the horizontal axis.) Thus if a preference for timber frame reduces demand by 30% the curve would move as in Fig. 3.2.

We can now expand the demand function to give a more complete picture. The demand function for bricks might read:

$$D\,b\,f(Pb, Ps, Pc, Y, T, Z)$$

where:

Db = quantity demanded of bricks
Pb = price of bricks
Ps = price of substitute goods
Pc = price of complementary goods
Y = buyers' incomes
T = tastes, fashion
Z = all other relevant factors.

NB: Any change in the **price** of bricks will lead to a movement along the demand curve; any **other changes** in demand will cause the whole curve to be shifted.

SUPPLY

Supply refers to the quantity of goods suppliers are willing and able to offer for sale in the market. This is not necessarily the same as the amount being

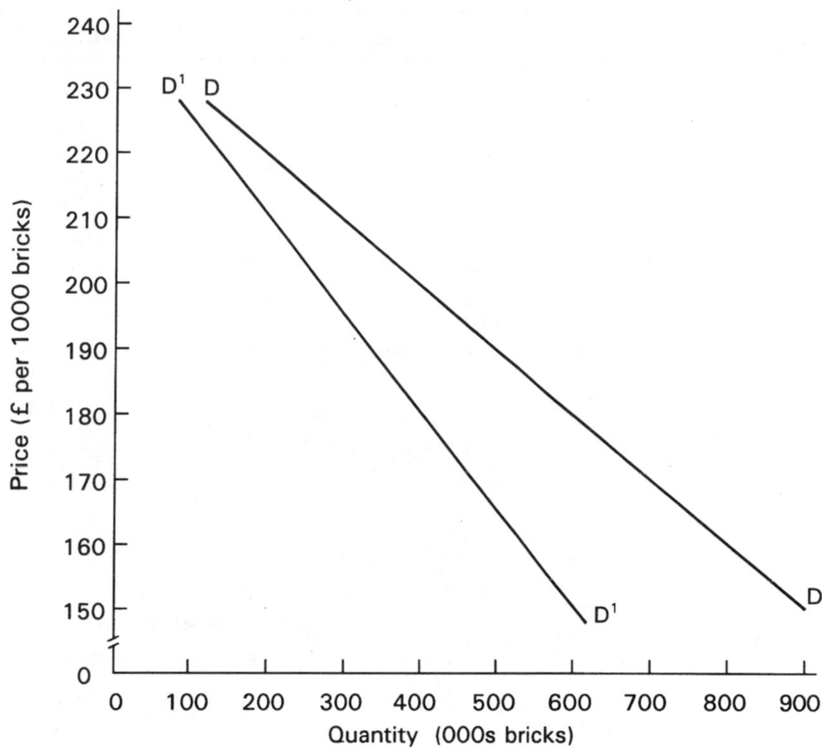

Figure 3.2
Decrease in demand (curve shifts left).

produced, as there may be times when stockpiles are being built up or run down, or when the price is insufficient to cover the costs of marketing the goods although they have been produced (as when crops are ploughed into the ground instead of being harvested). Again we must separate the effects of price changes from changes in the market conditions.

As the price rises, suppliers can cover additional costs and increase profit margins. This encourages them to extend the quantity supplied. There is a positive correlation between price and quantity; as price rises, a bigger quantity is sent to market. We can see this in construction markets: when there is a boom in demand tender prices rise and firms take on more contracts. Higher prices enable builders to pay more money for the additional skilled labour they need to meet the demand; high prices also encourage more firms into the market. Returning to our example of bricks, the relationship between price and quantity can be expressed either as a table of figures (Table 3.2) or a graph (Fig. 3.3).

A high price for bricks gives brick producers an incentive to work overtime. If the price remains high, they will think about increasing their investment in production facilities. Exports can be diverted to the domestic market, firms may even start importing bricks. Thus the quantity offered

Table 3.2
Hypothetical supply schedule: weekly supply of bricks

Price (£ per 1000)	Quantity ('000s)
230	1000
220	920
210	840
200	760
190	680
180	600
170	520
160	440
150	360

for sale is increased. But when the price is low, kilns may be taken out of production and workers laid off. In short, supply is a function of price:

$$S f P$$

Supplies, like demand, are also subject to non-price changes. Supply might be affected by weather conditions, new technology, methods of production, labour disputes, transport facilities, etc. For instance, the development of 'self-firing' bricks, which used less fuel, meant more bricks

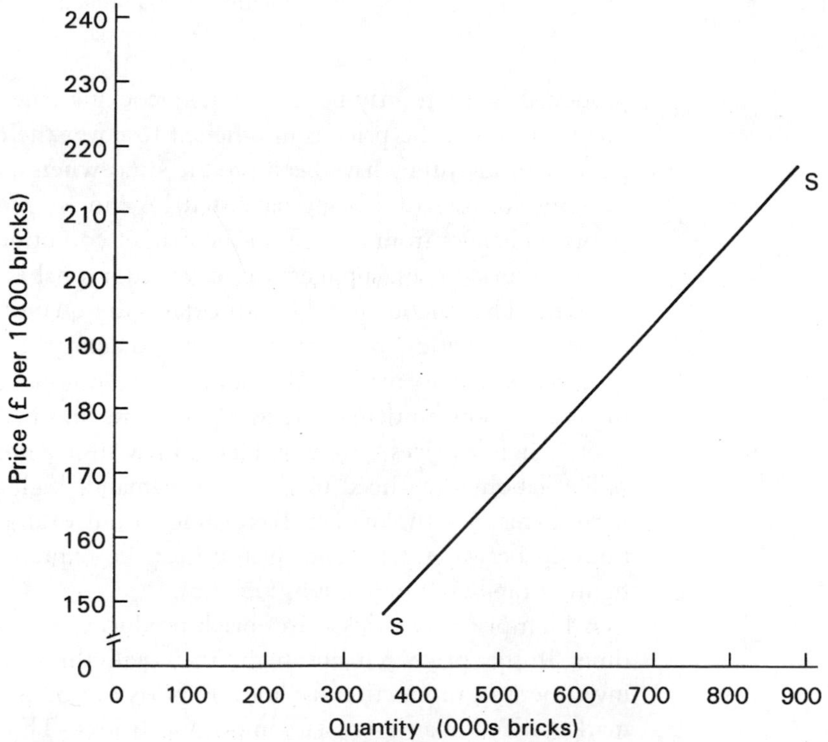

Figure 3.3
Supply curve (000 bricks).

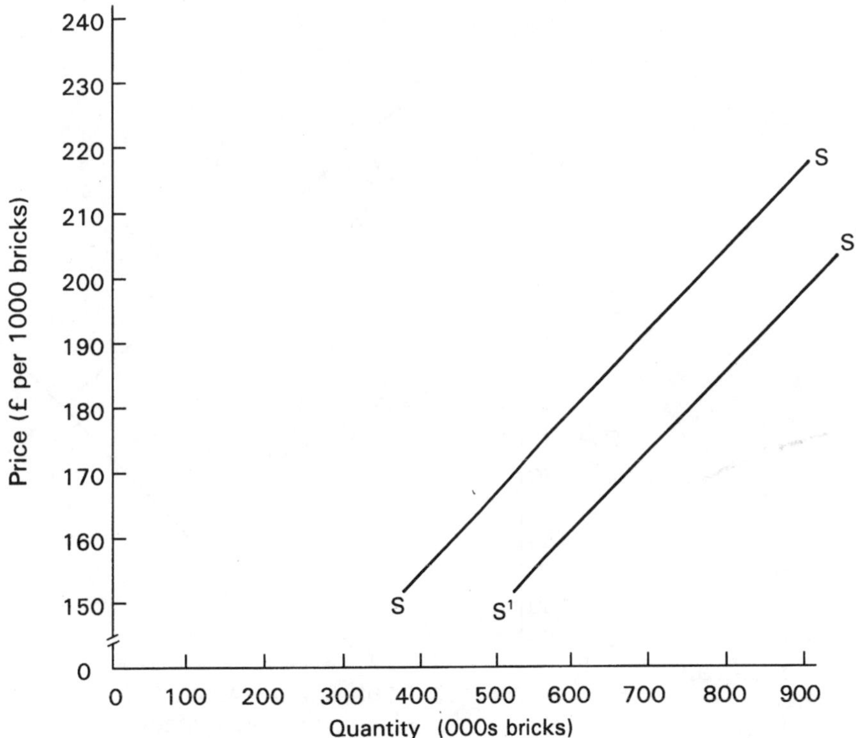

Figure 3.4
Increase in supply (curve shifts right).

could be produced and supplied to the market for the same cost as before. Here the supply curve shifts to the right, with an additional 150 000 bricks being supplied at each price level (Fig. 3.4).

PRICE DETERMINATION

When supply and demand are brought together it is clear that there is only one price level where consumers' buying intentions coincide exactly with suppliers' sales plans. In a perfect market the price will tend towards this point, i.e. where the two curves intersect (Fig. 3.5). This is the **equilibrium**, where supply and demand are perfectly balanced. At any other price level there is disequilibrium and pressure on prices to change.

Too high a price, e.g. £200, attracts more supply, but discourages consumers. An excess stock of goods builds up, so to get rid of it sellers must cut the price. Too low a price, e.g. £170, will mean excess demand and not enough bricks to meet all the orders. As stocks run out delivery times lengthen and builders suffer delays. The shortages will result in prices being raised as some buyers will pay more to get what they need. Eventually an equilibrium is reached, in our example at a price of £180. Brick yards have

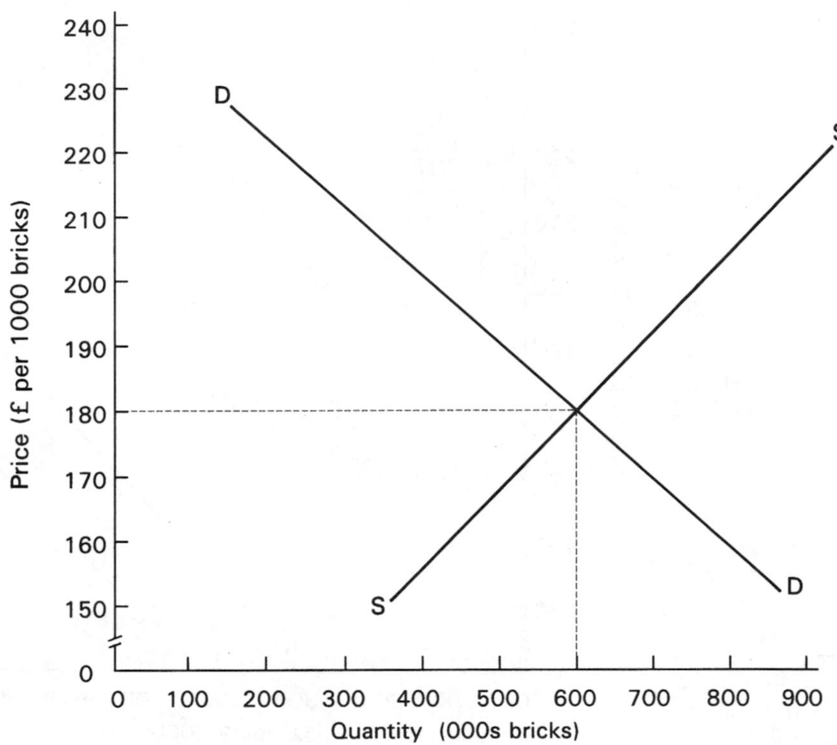

Figure 3.5 Equilibrium price and quantity.

no surplus stocks and customers can get all the bricks they want at that price.

Once equilibrium has been achieved there is no reason for further movement. The market has reached a point of stability, at least until either consumers or suppliers alter their plans – in other words until there is a shift in the position of the demand (or supply) curve. This creates an imbalance, either a shortage or surplus of bricks at the existing price. Consequently the price changes until a new equilibrium is established.

If, for example, a successful advertising campaign on the theme 'Bricks are Beautiful' increases demand (D curve shifts to the right) there will be insufficient supplies at the current price of £180. Prices will start to rise, reflecting the shortage. Suppliers respond to higher prices by increasing output (shown by moving up the supply curve) until, eventually, supply and demand are equal again. A new equilibrium is established, with more goods being bought at a higher price, reflecting consumers' greater desire for the product. An initial reduction in supply (S curve shifts left) would also create a shortage and cause the price to rise. Sales would fall, not because buyers wants have changed – the position of the demand curve is unaltered – but because they cannot afford to use bricks so freely at the higher price (move upwards along the demand curve).

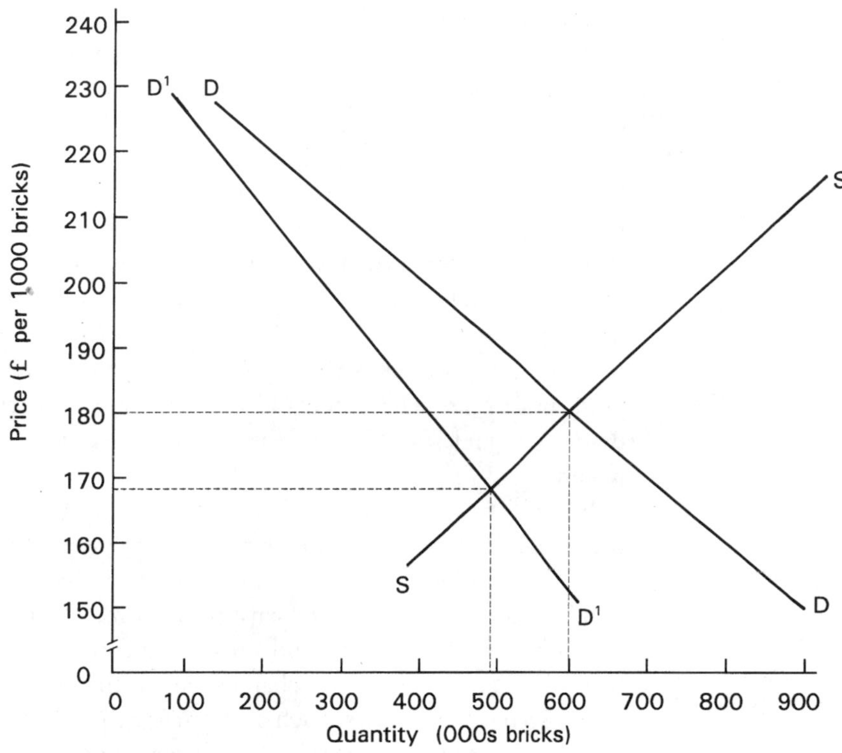

Figure 3.6
Effects of a decrease in demand on equilibrium price and quantity.

The effect of a decrease in demand is shown in Fig. 3.6. A change in conditions of demand, such as a fall in incomes, might cause this to happen. The result will be excess supply until suppliers adjust their prices and quantities to restore equilibrium. The effect of an increase in the quantity of bricks supplied can be analysed in a similar manner – you may like to try working out this diagram for yourself.

PRICE ELASTICITY

We know that price affects the quantity traded. If increased demand pushes up prices sellers strive to supply more goods. If a cut in supplies forces the price up, consumers will buy less. The effect on sales may be quite small or it may be substantial. **Elasticity** measures the degree of response to price changes. The more price sensitive we are, as consumers or suppliers, the greater the elasticity of demand or supply. Elasticity is calculated according to the formula:

$$E = \frac{\text{\% change in Quantity}}{\text{\% change in Price}}$$

If the result is greater than 1, e.g. a 5% price change produces a 7% change in the quantity traded, the response to the price change is elastic. If the result is less than 1, e.g. a 5% price change produces only 4% change in quantity, the response is inelastic.

Price elasticity has an important bearing on firms' revenues. These depend on both the quantity of sales and the price. If higher prices reduce sales, will revenue rise or fall? (Total revenue = Sales × Price.) The answer depends on elasticity. If buyers are not price sensitive (demand is inelastic) sales will fall very little, so revenue rises because each sale brings in more money. If demand is elastic, however, less money is spent on the product as the higher price is more than offset by fewer sales. With price cuts, where demand is elastic, sales and revenues increase, but if demand is inelastic sales increase so little that the lower price means less revenue.

Elasticity is greater when there are plenty of good substitutes available to the consumer. Hence the fierce price competition between different brands of similar products – a small price cut can induce many people to switch brands. People are also more sensitive to price changes which they perceive as significant; this will depend on their income level, patterns of expenditure, the size of the price change, etc. Demand is generally inelastic for goods seen as necessities, such as electricity, just because consumers do not feel they have much choice as to whether or not to buy them.

Elasticity of supply is measured in the same way as elasticity of demand, using the percentage change in price and quantity supplied. There is one additional consideration – the time it takes to adjust to new price levels. Buyers can often respond quite quickly to a price change, but suppliers may find it necessary to reorganize production, which takes time. Unless there are stocks to draw upon, supply is unlikely to be able to respond straight away. This is the **immediate or momentary** time period, during which supply is almost totally inelastic or fixed. Any shift in demand is reflected only in price changes, so a sudden increase in demand will create a shortage and push up the price. Given a little time to adjust, suppliers will make what changes they can within their existing production capacity. They may be able to put in an extra shift, or hire more workers to increase output. Supply begins to respond and price eases a little from the previous high level. This is the **short-run** position. In the **long-term** firms can expand or contract their capacity. They can build new plant or shut down some of their factories. Firms can enter or leave the industry. The more time allowed, the more elastic the supply curve becomes, as the industry responds to the situation signalled by price changes.

The diagram in Fig. 3.7 shows how supply responds to a change in demand from D to D^1.

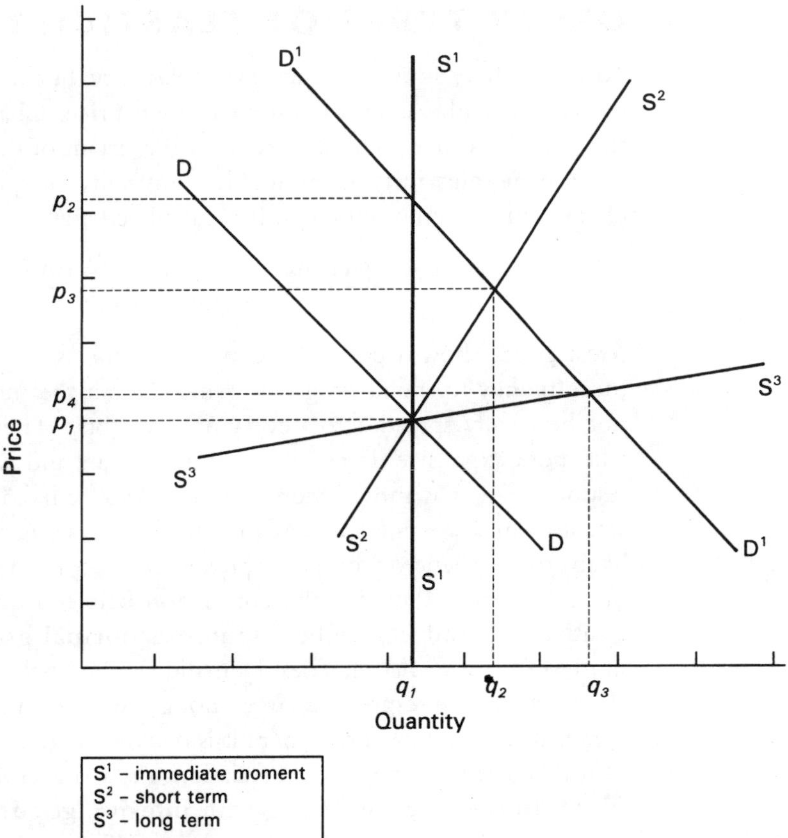

Figure 3.7
Elasticity of supply over time.

S1 – in the absence of stocks supply remains unchanged (totally inelastic) in the immediate moment. A shortage results and prices rise from p_1 to p_2.

S2 – over the short term the price rise encourages extra output from existing capacity. Pressure on price eases (p_2 to p_3) as quantity supplied increases from q_1 to q_2.

S3 – in the longer term new capacity can be created and the industry expands. Output increases to q_3 and price settles at a new equilibrium of p_4.

In the very long run it is difficult to separate the effects of price changes from other changes that are bound to occur in the economic environment. An initial fall in the price of houses because buyers are unable to afford mortgages when interest rates rise may be quite easy to discern. The longer-term effect on house sales and prices is going to be less easily distinguished because the market is also affected by wage rates, employment levels, etc., which are all changing in the longer period.

OTHER TYPES OF ELASTICITY

So far we have looked at own-price elasticity, but demand can be affected by variables other than the product's own price. Changes in income or in the prices of other goods are the most important of these.

Income elasticity measures the sensitivity of quantities demanded to changes in consumer incomes. It is calculated as:

$$\frac{\%\text{ change in Quantity demanded}}{\%\text{ change in Income}}$$

Most goods show a positive income elasticity, as a rise in income gives us more money to spend on goods generally, but the increase in demand will not be evenly spread across everything we buy. The demand for luxuries and more expensive items, such as houses, cars and yachts, shows a greater response, i.e. it is more income elastic, than demand for minor, mundane articles which are purchased as a matter of course. A rise in income is more likely to be spent on home improvements or a new car than on buying more groceries. Provided the correlation between quantity and income is positive, the products can be described as **normal goods**, that is we spend more on them as income rises. Demand can be further described as income elastic where the response is large enough for the elasticity coefficient to be bigger than 1 and inelastic where it is between 0 and 1.

Where there is a negative correlation, i.e. a rise in income brings about a fall in demand, we are dealing with **inferior goods**. The term 'inferior' does not necessarily imply poor quality. What it does mean is that consumers prefer some alternative, which they could not afford before, but can afford when their incomes rise. In the 1950s and 1960s most British houses had open fires. In the 1960s and 1970s, as incomes rose, people wanted more comfort and central heating systems became the norm. Open fires came to be regarded as inferior goods, to be replaced if possible. As income rose the demand for open fires fell. Nowadays many people are once again installing open fires in their homes, not as the primary source of heating but as an addition to give the extra cheer of a real fire. At one time normal goods, they came to be seen as inferior goods, and are now once again normal, even luxury goods!

Cross-price elasticity looks at the way in which the quantity demanded of one product responds to a change in the price of a different product. Complementary goods or close substitutes are most likely to show a strong correlation. The formula here is:

$$\frac{\%\text{ change in Q demanded of A}}{\%\text{ change in P of B}}$$

Thus, people might buy more Apples because Bananas had gone up in price. In the housing market demand is particularly sensitive to changes in the price of credit (mortgage interest rates) because the majority of houses are purchased with borrowed funds. Home buyers are buying a package consisting of a house, costing £*x*000, plus a loan, costing *y*%. If interest rates were to rise from 12% to 13% (a change of 8.5% in mortgage prices) and as a result the demand for property fell by 10% the housing market would be described as elastic with respect to interest rates. In this case the goods are complementary, so a rise in the price of one causes less demand for the other, a negative correlation. Where goods are substitutes, the correlation will be positive; if the price of renting accommodation fell, more people would rent and fewer would buy, so the demand for owner-occupied property would also fall, *ceteris paribus*.

A FURTHER NOTE ON CALCULATING ELASTICITY

$$\text{Price elasticity} = \frac{\%\text{ change Q}}{\%\text{ change P}}$$

Referring back to the demand schedule for bricks, as the price moved from £200 to £190 per thousand demand altered from 400 thousand to 500 thousand. The changes of £10 and 100 respectively must be expressed as percentages before they can be compared and the value of elasticity can be calculated. Where the change involved is large, as from 400 to 500, the percentage is affected by which figure is chosen as the base. For example, 100 = 20% of 500, but it is 25% of 400. If the change is very small this variation can be ignored, but where it is sufficient to distort the comparison, the calculation should be based on the mean (midway) price or quantity, for example 100 as a percentage of 450 and £10 as a percentage of £195. This will eliminate the inconsistency of different values for elasticity according to whether it is calculated for a price rise or a price fall.

Understanding elasticity is very important for companies developing their commercial strategies and formulating pricing policies. Income elasticity and cross-price elasticity can help to forecast market trends. As society grows richer we look for better quality, higher specification goods rather than cheap, basic goods, and we spend more of our incomes on housing and consumer durables. In the shorter term cross-price elasticity and own-price elasticity are important in judging buyers' reactions to price changes by firms and their competitors. If Cosy Cat Homes are offering first time buyers a 10% discount on their Greenfields estate, how will it affect sales of their rival, Oak Tree Homes on the neighbouring Meadowlands development?

CASE STUDIES

A. Good times ...

The following extracts are loosely based on various news items which appeared during 1988.

Wales has felt neglected by the developers for years but now the market has burst into life, fuelled by the M4 corridor and talk of economic recovery. Alan Jones estate agency says house prices have leapt by 30% in the last quarter and Cardiff is booming. Around 25% of the buyers flocking in come from places such as London, Bristol, Birmingham, Southampton, Manchester and even Glasgow. The Japanese, unfettered by the preconceived ideas the English have about the place, got in there long ago.

Wales used to say 'come here and we will give you grants', but now it's 'come to Wales for a better lifestyle'. Travel to work time is short, and the sea and the mountains are 20 minutes away ...

Britain's current house price boom has not actually reached the status of a mania, but it incorporates its share of madness; people scrambling to buy houses at ever higher prices in the profound conviction that prices will rise forever, and going up to their ears in debt to back this belief.

In 1979 the value of personal homes was £270.6 bn. It is now (1988) around £880 bn, a rise of 225%, but only £75 bn worth of new homes have been built in this time ...

If telecommuting catches on, employers would need to provide less office space in the south-east, they could drop London weighting and company cars, and might experience up to 33% increase in productivity.

Housebuyers would look for more rooms in a house (old vicarages would be ideal), better sound insulation and more sophisticated heating controls to create the right office atmosphere. There would be less need for key workers in the big cities. The City alone could lose 113,000 service workers to residential areas. More people would be needed in the suburbs to bring in take-away meals, clean houses and deliver the teleshopping ...

B. ... and bad

Redland, one of Britain's biggest building materials manufacturers, is to close three brickworks in southern England with the loss of about 200 jobs. The closures will lead to a cut of about one third in the company's production of bricks. Stocks of unsold bricks held by manufacturers have risen to more than 1 bn, the highest level since 1982. About 60% of bricks are used in house building. Prices of standard housing bricks have fallen by approximately 25% in the past year.

WORKSHOP

NB: This chapter has used the model of perfect competition to explore the price mechanism through the example of bricks. The ordinary fletton house building brick matches two of the requirements for a perfect market: it is a reasonably standard (homogeneous) item and widely used by many builders around the country. In other respects it is a less good match. Brick supply is dominated by a few major producers who have considerable influence on prices and output. As a result it does not always respond very quickly to changes in the market.

Construction markets often react slowly, and one change can be overlaid by many others before the market has finished adjusting to the first change. The assumption of *ceteris paribus* is therefore unlikely to be met. It is for this reason that we use a model, the perfectly competitive market, to examine how prices work. In the real world most markets are imperfect, but the forces of supply and demand still operate in the ways analysed above.

1 Compare the similarities and differences between the housing market and the model of a perfect market.

2 Using the figures given in the text for the market in bricks calculate the value of E for a price change from:
a £190 to £180;
b £160 to £150.
Is demand elastic or inelastic in each case? How do you account for the difference between the two cases?

3 What are the conditions of demand which made the market in Wales (case study **A**) so buoyant in 1988?
Illustrate the impact on house prices in south Wales with a supply (S) and demand (D) diagram.

4 The extracts in case study **A** suggest higher house prices were accompanied by greater demand. Economic theory states that the quantity demanded is usually inversely related to price changes. How do you explain this apparent contradiction? (Try using diagrams to help you analyse the situation. Remember to distinguish between a move along the curve and a shift in the position of the curve.)

5 The effects of the computer on our lifestyles has major implications for construction markets. Using diagrams, consider the effects on:
a regional housing markets;
b demand for older *v.* newly built property;
c office development. (Case study **A**)

6*The housing market is often said to be price inelastic but income elastic. Explain what this means and discuss why it should be so.

7 Define the term 'market equilibrium'. What evidence is there to indicate whether or not the market for bricks was in equilibrium? (Case study **B**)

8 Draw a diagram to illustrate what had happened in the brick market. Give a brief explanation. (Case study **B**)

9 Define the term 'elasticity of supply'. Calculate the elasticity of the response shown by Redland and say whether this was elastic or inelastic. (Case study **B**)

DISCUSSION QUESTION

❑ In view of the somewhat 'unreal' assumptions of the perfectly competitive market, is this model worth studying if you want to understand markets in the 'real world', such as the housing market?

4 Factor markets: land

PREVIEW

- What are the resources needed to build houses?
- Who buys these resources?
- Does it make any difference whether the buyer is a speculative builder, or someone building their own home?
- What determines the price of houses?
- What is the relationship between the price of houses and the price of land?

DEMAND FOR FACTORS OF PRODUCTION

Consumer goods and services are purchased for themselves, because they satisfy a want or need. Provided they offer us enough utility we will pay the market price, at which the goods are worth supplying. Factors of production, the resources of land, labour and capital, are not bought for the direct satisfaction of wants in this way. The buyer is an entrepreneur, not the end user but a businessperson who intends to use the resources for production to generate profits.

The resources themselves are only a means to an end. Demand for them is indirect, it is a **derived demand**, i.e. it is dependent on there being a prior demand for the goods or services which are going to be produced. The link between factor prices and demand for factors of production is thus less direct than the link between consumer goods and prices. The consumer can weigh the cost of product against the satisfaction obtained: if it is 'worth' the price, it is worth buying. The producer must judge the output that can be achieved and whether it can be sold at a price that will cover the cost of the resources used to produce it.

Bearing in mind that we are dealing with a derived demand, we can now apply supply and demand analysis to factor markets. Of course market forces do not operate with total freedom in any modern economy and factor markets are often subject to external regulation, e.g. planning controls, labour laws, the manipulation of interest rates, etc. As a result prices are not always able to respond fully to market pressures and the market may remain in disequilibrium for a considerable time, with shortages or excess supply uncorrected by price adjustments.

ECONOMIC CHARACTERISTICS OF LAND

Land, as a factor of production, refers to all natural resources unimproved by human work. A piece of land in the usual sense, e.g. a field, may combine the original gift of nature (a portion of the earth's surface) with the results of human effort (to clear, till and improve the fertility of the soil) and the market value will reflect any improvements or damage. Its price can be expressed either as an outright purchase price or as an annual rent, either of which will reflect the earning capacity of the land. If the elements of labour and investment in any man-made improvements are set aside, what is left is a price for nature's gift, a piece of land.

It has cost nothing to produce and it exists, regardless of what it is used for, or whether it is used at all. Its price is not due to the costs of production, which are zero, but depends on the level of demand and the limited supply. Since the amount of land on the earth's surface remains virtually fixed, its scarcity value depends on how much demand there is. As our need for space grows, so the land, which remains the same, becomes more and more scarce.

THEORY OF RENT

These unique characteristics of land – zero production costs and a fixed supply – gave rise to the theory of economic rent, developed by Ricardo in the early nineteenth century, a theory which is still important to our understanding of factor prices today. Ricardo argued that the payment, or rent, charged for a piece of land had no influence on the amount of land available. The supply of land was dependent on nature and the owner incurred no extra cost by allowing another to cultivate it. Any rent received was, therefore, pure profit. How big a profit would depend entirely on how much rent tenants were prepared to pay – on the demand for the land – which in turn would depend on how much could be earned by farming it. If crop prices rose, landlords creamed off the profits by raising rents.

Rent was seen to be a surplus profit, not an essential cost of production. This sort of payment is called **economic rent** to distinguish it from the ordinary sense of rent as a hire charge. Economic rent does not arise out of the expenses of production. (In the case of the field it can be argued that the landowner had to pay to acquire the land, but this is not a production cost – the land would still have existed had it not been purchased. The buying price is simply a transfer fee, paid to obtain a change of ownership.) Payments like this, which arise out of scarcity rather than production costs, are not restricted to land. Labour or capital may be able to earn economic rent also.

Although allowing the land to be farmed, or built on, or used in some other way does not involve the landowner in any production cost, there is still an opportunity cost to consider. There are generally alternative uses for land, each with a different revenue potential. In this case there is a cost in putting it to any one use, namely the loss of earnings from the best alternative use. In these circumstances it would be wrong to describe all the earnings as pure profit or economic rent.

Describing rent as pure profit implies the landlord would otherwise be getting nothing, but if the land has another use it can still bring in some revenue. Only the excess payment, over and above the next best alternative, can be termed economic rent. The same applies to all factors of production, not just land. To retain the use of land, labour or capital an entrepreneur must match the next best offer, otherwise the factor will be re-deployed to another use. The minimum payment which keeps the factor from going elsewhere is called its **transfer earnings**. Transfer earnings are not rent; they can be seen as costs because they are the lowest sum which will secure the use of the factor in question. They have to be met in order to continue production, but anything more can be considered rent, a premium due to scarcity.

In the case of land we can say that in total it is fixed in supply. It is a gift of nature and any profit from it is pure gain. The earnings to land as a whole can thus be classified as rent. For a particular piece of land, however, this analysis is too simple. Although the total land supply is fixed, land for building, or for growing food, or for recreation, is not. Land can be transferred from one use to another. If the land is currently employed in one way anyone wishing to acquire it for another use must be prepared to pay an amount at least equal to its current worth. This is its transfer earnings or opportunity cost. This is not a surplus, but a cost of production in respect of what the land is to be used for. It is only earnings in excess of this level that can be seen as rent (Fig. 4.1).

Opportunities to earn rent arise out of scarcity, whether natural or man-made. A farmer with fertile land will achieve higher crop yields per acre than his neighbour on poorer ground. The good land is more desirable, and because it is naturally scarce and will earn more revenue, people will pay more for it than for the poorer land; it earns economic rent. Land at the margin of cultivation produces just enough crops to repay cultivation. If

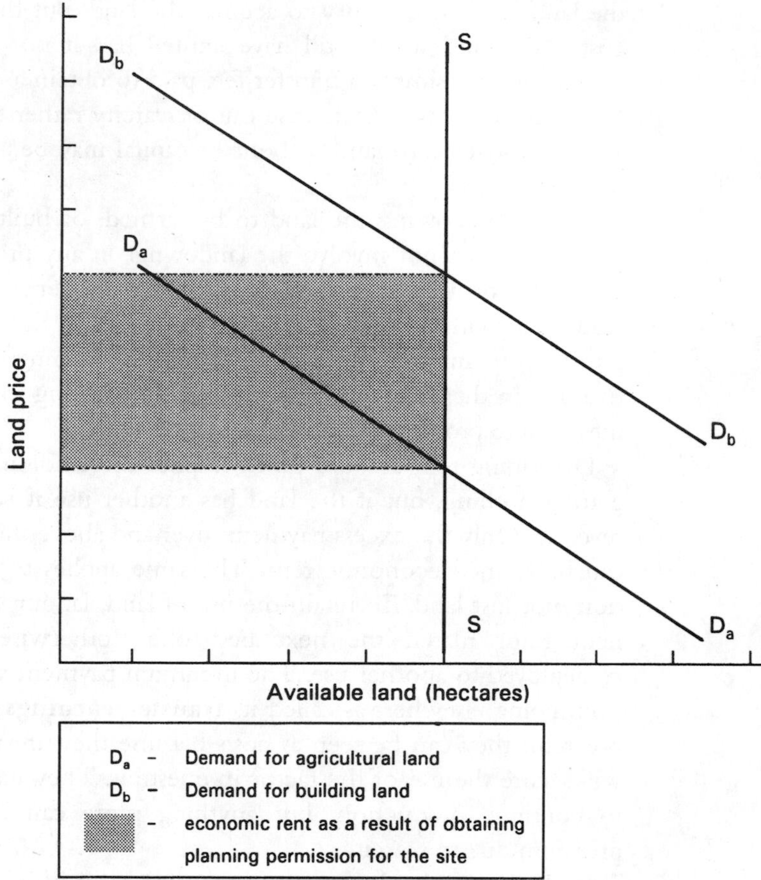

Figure 4.1
Economic rent.

crop prices fall poorer land will go out of cultivation unless the farmer can find other ways to supplement the farm income. (Table 4.1 provides some examples of agricultural land prices on the market in April 1993.)

Similarly, a developer who wants to build will earn more profit if his land is chalk, which offers a firm base for building, than another developer with sites on shrinkable clay or running sand, where the costs of construction are higher, leaving less profit. If both earn enough to make their projects worthwhile the chalk-based developer stands to make a surplus, or economic rent. However, the owners of either site will profit even more if the land is re-zoned for housing rather than low grade industrial development. This creates a windfall gain or abnormal profit (economic rent) because there is greater demand for housing than industrial property.

Planning restrictions which limit the supply of land for certain uses increase the scarcity of that category of land. The supply of building land becomes less elastic where green-belt restrictions limit development. As

Table 4.1 Agricultural property prices (April 1993)

Newton Farm £1000 per acre	Fife in Scotland. Close to the Edinburgh–London railway and St Andrews. 494 acres: trout loch, clay pigeon shoot, off-road 4×4 driving course.
Cardwell £2300 per acre	Holbeach in Lincolnshire. Tenanted farm: 1st grade silt land; cereals and vegetables.
Bron-Rhydd £867 per acre	Powys, Wales. 530 acres: hill farm with sheep and cattle.
Stream Farm £2100 per acre	Taunton in Somerset. 200+ acres: mixed farming, fishing lake stocked with rainbow trout.
Bishopstoke £751 per acre	Near Grimsby. Tenanted farm: 599 acres, 2–3rd grade land, arable crops.

demand for housing space increases this will lead to house prices rising faster in the green belt than outside (Fig. 4.2). Homeowners enjoy a capital gain. Land prices in the green belt remain low because there is less demand for land which has no prospect of development. Sites which do have planning permission, however, will increase in value because of their scarcity.

It is worth noting here that the often expressed view that houses are expensive because land prices are high is mistaken. The demand for building plots is derived from the demand for housing. If house prices rise builders want to build more houses, which pushes up prices of building land. So why should house prices rise? Not because it costs more to build since most houses on the market are existing properties; new houses add only 1 or 2% p.a. to the total housing stock. Current land prices and construction costs clearly do not affect the production cost of older properties. Where supply is inelastic (fixed) prices are driven by demand. As the housing stock changes only slowly, it is demand that sets the pace.

New houses have to be competitive with prices for older houses. Buyers may be willing to pay some extra for a brand new property, with all the latest features and newly decorated throughout, but they will not pay prices that are quite out of line with the rest of the housing market. New and older houses are close substitutes and price changes will be reflected across both types of property.

Builders must therefore estimate the sale price in the current market, deduct the costs of construction, including their own profit, and the residual sum remaining is what is left for buying the land. Where house prices are high, builders will be able to pay more, and because they compete for plots land prices will be pushed up. A building plot in the South East may

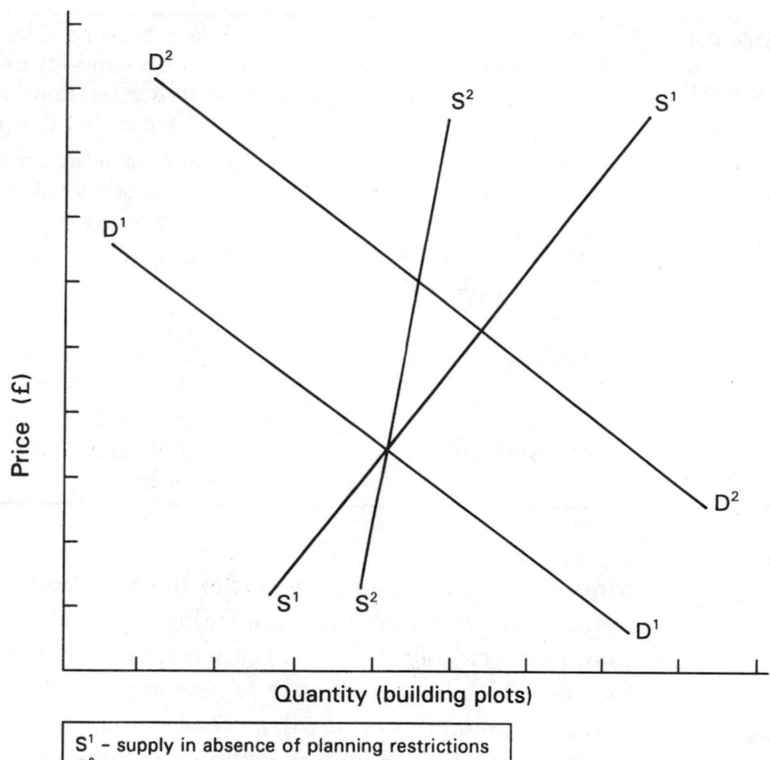

Figure 4.2
Supply of building land restricted by green belt.

be intrinsically no different from a plot in the North West, but if there is more demand to live in the South East prices of housing and land will be higher there – and opportunities to earn economic rent greater.

The theory of rent is not restricted to land. One more example may help to clarify it. Suppose a professional sportsperson sets a target of £20 000 income, with the intention of retiring from the sport unless this is achieved. If the athlete then wins £25 000 in the next major event he or she enters, earnings now exceed the minimum required by £5000. This surplus may be termed economic rent, because it is more than was needed to keep that person in the sport.

Building workers can also earn a surplus; in a boom it is difficult to find enough skilled labour to do the work available. Firms will pay well above the minimum rates and good tradespeople can earn a substantial element of rent. (The downside is that they may find it hard to earn even the standard rates when there is a recession. Instead of a shortage of labour, the industry then has a shortage of work and those whose incomes fall below their transfer earnings will seek better paid jobs elsewhere – if they can find them.)

Similarly capital and enterprise can also earn rent when returns are greater than required to ensure that they are not re-deployed in some alternative use. At the top of the boom tender prices rise, not only to cover the higher costs of labour just mentioned, but also because profit margins are increased; firms are earning an element of rent. When there is less work available they will have to reduce prices. This sort of rent, which is only temporary, is sometimes called **quasi-rent** to distinguish it from the more permanent surplus which can be earned where factors are in fixed supply.

LAW OF DIMINISHING MARGINAL RETURNS

Another economic theory which owes its origin to the fixed nature of land supply is the law of **diminishing marginal returns**, or **variable proportions**. This shows that the relationship between inputs and output depends on the proportions in which the inputs, or factors of production, are combined. If one factor is fixed in quantity (e.g. land) and this is combined with successive units of a variable factor (e.g. labour), we would expect output to increase as more labour is employed. Initially the increase in output may even grow faster than the increase in the labour force, but there comes a point where further additions of labour, the variable factor, produce successively smaller increments to output. Growth in output is slowing down – we are experiencing diminishing marginal returns.

Suppose, for instance, that to speed up the rate of progress (output) on a building site (fixed factor) the number of bricklayers (variable factor) is increased. At first a 10% increase in the labour force might result in a similar increase in the number of bricks laid per day, but further additions to the labour force will not continue to produce the same pro rata improvement in output. There is limited space on site for the bricklayers to work, they will start to get in each other's way, bottlenecks will occur as sections of brickwork are completed and they have to wait for other tradespeople to complete their tasks. In theory the variable factor could be increased to the point where marginal returns became negative, i.e. there are so many bricklayers elbow to elbow that they have not enough room to work and total output begins to fall. In practice we would stop employing bricklayers long before that point was reached.

The law of diminishing marginal returns was developed at a time when Britain was experiencing its first population 'explosion', on the eve of the industrial revolution. Thomas Malthus (1766–1834) was struck by the contrast between humanity's unrestricted potential for reproduction and our island's limited space and restricted potential for growing food. His gloomy conclusion was that periods of plenty would result in rapid population growth which, unchecked, would soon outstrip the available resources. The end would be inevitable poverty and hunger. The population would then be decimated by famine, disease or war, all consequences of the population pressures – and the cycle would start again.

Today our population is ten times what it was in Malthus' time, yet our prosperity would have been unimaginable by his contemporaries. How have we avoided the cycle of misery which he predicted? Malthus developed his ideas at a time when fundamental economic changes were under way. He did not fully realize the potential of the New World and international trade, which loosened the limitations of our island boundaries and enabled us to draw on the resources from around the world. In effect we increased the size of the fixed factor (land) by using food grown on far off lands. Nor did he foresee the immense changes in technology and methods of production which transformed manufacturing and farming and greatly expanded our wealth. The law of diminishing marginal returns applies to a given situation, *ceteris paribus*. If the fixed factor is enlarged, or if methods of production are changed, for instance new technology is developed, the relationship between inputs and output is altered and the analysis must start again from a new base line.

CASE STUDIES

A.

Eagle Star Properties put forward proposals for a new market town to be developed in north Hampshire in the early 1990s. The plans centred round Micheldever Station on the main railway line from London to Southampton and Bournemouth.

The company, which owns 10,000 acres of predominantly agricultural land in the area, argued that a proper community of 5,000 homes was a better way of meeting Hampshire's housing needs that the traditional 'cramming and infilling'. They sought to develop 550 acres at Micheldever Station, to include 49 acres of town centre, with retail space and public buildings, and a further 45 acres of light industrial and distribution space. The town would ultimately occupy over 1,000 acres.

It was hoped that planning permission could be granted in 1993, site work commence in 1996, and the town reach completion by 2010. The development would take place in four phases, the first of which would include 10 acres of employment land and a quarter of the planned residential development.

In 1993 the Government announced the rejection of this, and a number of other, similar schemes.

B.

In 1989 the city of Bath was shocked by the news that one of its oldest manufacturing firms was to be broken up and sold. Stothert and Pitt had a worldwide reputation for the quality of their cranes and other engineering products. The contractor's plant division had been responsible for significant advances, including the introduction of a vibrating roller. After two hundred years in the business, its closure by the parent company, Hollis Industries, was justified on the grounds that the sale of Stothert as a number of separate businesses, together with the sale of its site, was needed to repay bank debts.

Stothert's Victoria Works was situated near the city centre, in a prime development site of 15 acres. Estimates of the land's value ranged from £4.5 m to £15 m depending on the type of development permitted. Bath City Council were anxious that the land should continue to be used for industrial production in view of the small number of sites available to industry in the city. Developers argued the site would be given greater value if housing and offices were allowed.

For Stothert high land prices put it in a position where the increasing value of its site, far from being an asset, led to its downfall. If the company had occupied a less valuable site, in the depressed north-east rather than the prosperous south-west, it might not have found itself put up for sale.

WORKSHOP

1 Referring to Table 4.1 and using the concept of economic rent, can you explain why the prices of agricultural land vary so widely?

2 With the help of a diagram, explain why Stothert (case study **B**) were unable to continue their business manufacturing construction plant at the Victoria Works in Bath. Assume that the lowest price quoted for the site was just sufficient to make redevelopment more profitable than continuing the present business. Show on your diagram:
a the transfer earnings
b the economic rent
if the site were given planning permission for housing and commercial developments.

3 Explain, using diagrams, the effects on the supply elasticity and price of:
a building land
b agricultural land
if the Eagle Star proposals (case study **A**) had been accepted.

4 If planning permission had been granted, in case study **A**, who was likely to gain and who was likely to lose as a result of the new development?

5* 'Cheap land reduces building costs and keeps down property prices.' How valid is this argument?

DISCUSSION QUESTIONS

Giving planning permission rewards landowners for nothing more than being fortunate enough to own land! Where land values rise solely because of community decisions about the use of land, the rise in value should be returned to the community.

- ❑ How could this be achieved?
- ❑ Do you agree with the view stated?

5 Factor markets: labour

PREVIEW

- What determines the demand for labour?
- Can people 'price themselves out of jobs'?
- Would lower wages create more jobs?
- What determines the supply of labour?
- Do higher wages encourage people to work more hours?
- Will improved productivity lead to more or fewer jobs?

Labour is the contribution to production of human effort, physical or mental. The labour market has its own unique features because it involves people. People have their own motivations for working and the supply of labour is affected by these. As in other markets, price is an important element. We work primarily to earn a living, so financial rewards affect our willingness to offer our labour, but this is not the only factor.

LABOUR SUPPLY

The supply of labour depends on the amount of hours offered by each individual worker and on the number of workers available. Not everyone can choose how many hours they work, although self-employment and voluntary overtime allows some to vary their total hours. More generally hours are negotiated between the workforce as a group and employers, whether on a company or an industry basis. Changes in working hours and holiday entitlements will affect the amount of labour offered.

If the length of the working week is generally reduced, the supply of labour falls. Unless there are improvements in productivity this will mean we need more people in the workforce to keep up the supply of labour and maintain output. The wartime years saw the workforce expand rapidly as millions of women entered the labour market, often in roles such as welders and fitters which had previously been considered male preserves. After the war many of the women left the workplace and returned to the home but without them production could not have been maintained during the crucial wartime years.

In the longer term labour supply varies with changes in the population, its age structure and activity rates. Activity rates measure the proportion of the population of working age who are in work or available for work – these make up the labour force. Changes in the activity or participation rate allow the labour force to be more variable than population size. In recent years the activity rate for men has declined, as more school leavers opt for full-time higher education, which keeps them out of the workforce for two to three years. At the same time more older workers have opted for early retirement. By contrast female activity rates have increased as marriage and employment cease to be regarded as an either/or choice.

At present the labour force faces a period of shrinkage due to lower birth rates, which means that the numbers of school leavers entering employment in the 1990s may not be sufficient to replace those at the top of the age scale who are retiring. This implies a possible fall in living standards, unless compensated by other changes, such as an increase in activity rates (the proportion of the working age population who seek to work), a change in retirement age, or improvements in productivity.

EFFECTS OF WAGE RATES (PRICE OF LABOUR) ON LABOUR SUPPLY

How does labour supply respond to price? The price of labour is the rate per hour, or per week, that is offered. This is expressed in money terms but people are concerned about the purchasing power of their wage packet as well as its nominal (money) value. When prices rise they require an increase in money wages just to keep pace with inflation, not as an incentive to work more. The *ceteris paribus* assumption must be kept in mind when analysing wage rates. It is changes in real wages that affect the labour market. We should realize also that 'high' or 'low' wages are relative and depend partly on people's expectations.

At a moderately low wage rate labour supply is likely to respond to changes in price (wages) in the same way as other markets; higher wage rates make work more attractive and more labour is offered. This is partly because workers are willing to work longer, encouraged by the extra pay, partly because more people enter the labour market. Married women, for instance, may find the costs of child care make it impractical to work if wages are low. Choices between work, study or retirement will be influenced by wage levels.

As incomes rise and standards of living improve money becomes a less effective motivator. Having time to spend and enjoy their income conflicts with spending the time earning it. Labour supply is less responsive to wage rises. On a diagram (Fig. 5.1) the supply curve becomes vertical. It can

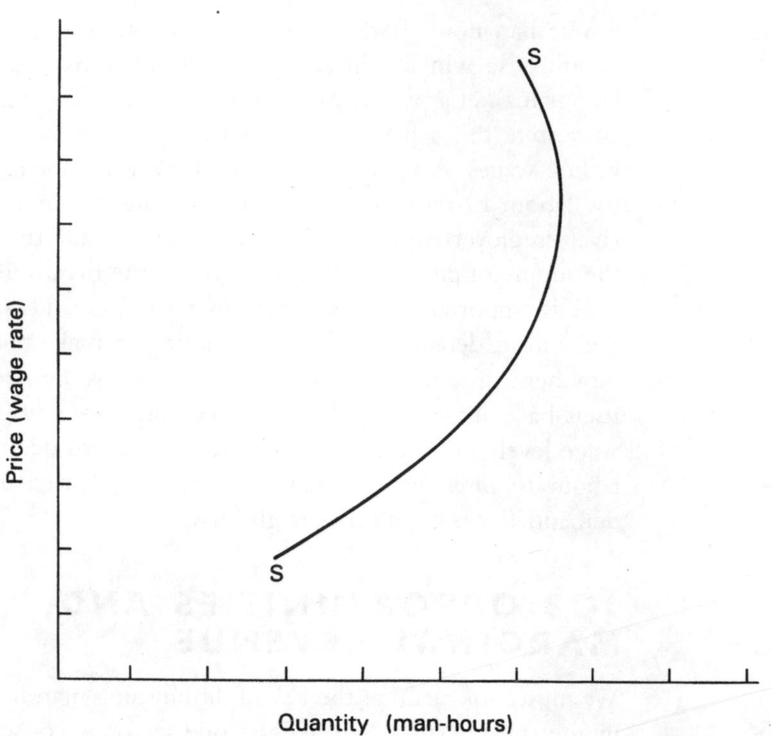

Figure 5.1 Backward sloping supply curve for labour.

even start to slope backwards at still higher wages. People have reached a position where a rise allows them to work fewer hours to earn the same amount of money, rather than work extra hours for yet more money. The self-employed building worker, when work is plentiful and the rates are high, may choose to leave the building site for the golf course on Friday afternoons!

At first the higher income and increased purchasing power made work more attractive as wages rose. This is the **substitution effect**, bringing about an increased supply of labour as people substitute work (and the goods it can buy) for leisure. But as incomes reach a level where the more urgent demands for goods are satisfied, the demand for leisure grows. The **income effect** of a wage rise is to increase demand for most goods, including leisure. As the income effect starts to outweigh the substitution effect the curve turns backwards. Extra earning power is buying more time off work rather than more goods.

Because the income and substitution effects work in opposite directions it is not easy to predict the outcome of a rise in wage rates. From a histori-cal perspective we can see that the workforce as a whole has traded s its potential income for extra leisure since wage rates started to rise s tially. In Victorian England working hours were longer and holi fewer than now. Trade unions today continue to work for better pay and conditions, which can include shorter hours. Like the individual, the labour market as whole may exhibit a backward sloping supply curve.

Despite this a firm that wants to expand its labour supply can do so by raising wages. A single firm, or industry, is competing with many others in the labour market. By offering higher rates it can attract labour away from rival employers. Individual workers may be able to reduce their hours but the supply of people willing to work for the firm is increased.

It is important to remember that the demand for labour is a derived demand: it depends on the firm's ability to make profits. The high unemployment of the 1980s led politicians to stress the need for labour to 'price itself back into the market' by accepting lower wage increases. Although wage levels are one factor in the entrepreneur's decisions about how much labour to hire, the correlation between the price of labour and quantity demanded is not entirely straightforward.

JOB OPPORTUNITIES AND MARGINAL REVENUE

We must look again at the law of diminishing marginal returns. This applies in the short term when at least one factor is constant. Assume the fixed factor is an area of forestry land. The entrepreneur must decide how much labour (the variable factor) to employ on it. The aim of the firm is to make a profit by selling timber. Physical output of timber is not an end in itself. To earn a profit the entrepreneur has to balance the change in costs as more labour is hired with the change in revenue. Revenue is a function of the quantity of timber produced and its selling price. This is shown in Table 5.1.

The first column shows the number of workers employed, increasing by one at a time. The second column gives the total yield of timber at each level of employment. From this it is possible to calculate how much extra timber is produced by hiring one extra worker. For example, if six workers produce 320 m^3 of timber and seven workers produce 360 m^3, adding a seventh worker has raised output by 40 m^3. This is the **marginal physical product** given in the third column. To find out how much this is worth to the firm it is multiplied by the price of timber, £3.00 per m^3, to give the **marginal revenue product**, i.e. adding a seventh worker raised the total revenue by £120.

When deciding how many men to employ the employer must compare the cost of labour with the revenue brought in. This can be done by calculating total labour costs and total revenues at all levels, or by comparing marginal costs and revenues. Most decisions are made at the **margin** – will

Table 5.1 Diminishing marginal returns

Workers	Total physical product (m^3)	Marginal physical product (m^3)	Marginal revenue product (£)
1	45	45	135
2	100	55	165
3	162	62	186
4	220	58	174
5	273	53	159
6	320	47	141
7	360	40	120
8	393	33	99
9	419	26	78

profits improve if one more (or one fewer) person is employed? If the extra cost is below the additional revenue it will pay to take on another worker. If we take the going wage rate to represent the marginal cost of hiring another person, then at a wage of £150 the employer will hire five workers. A sixth worker is not worth hiring, since the extra revenue brought in is less than the cost of the extra wage.

There are a number of ways in which changes in the situation could affect the number of people employed. Firstly a change in productivity (e.g. improved timber yields) will increase the volume of output and may mean that the sixth worker now produces enough timber to justify another wage. Secondly prices may alter so that the marginal revenue product changes even though output has remained constant. (Would a 10% rise in timber prices make it worthwhile taking on the sixth worker?) Lastly the marginal cost of labour may change. If wages rose by 6% the employer would just break even on the fifth employee; any greater increase in wages will mean laying him off, if profits are to be maximized. This is illustrated in Fig. 5.2.

Does this mean lower wages will produce more jobs as employers hire more people? Not necessarily. More labour produces more timber. If the extra timber can be sold without affecting prices all is well – falling wages will indeed lead to more jobs. The argument may not work so well if it is applied to the labour market as a whole, or even on an industry-wide basis. In this case the timber market may be over supplied and prices will fall. This lowers the marginal revenue product of labour and weakens the incentive to hire more people. The lower wages would have to be balanced against lower revenues.

The firm must also consider whether it can sell the extra output. How elastic is the demand for it (see Chapter 3)? If lower wages enable firms to

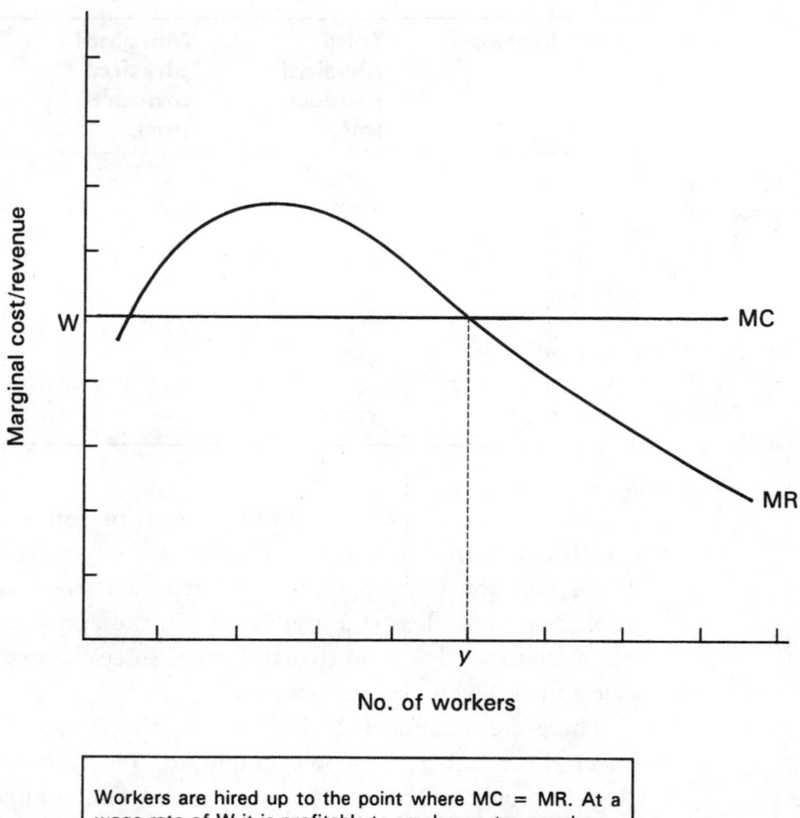

Workers are hired up to the point where MC = MR. At a wage rate of W it is profitable to employ up to *y* workers

Figure 5.2
Demand for labour.

reduce prices and if demand is elastic, then increased sales will mean that the firm's revenues rise. In this case lower wages are indeed likely to lead to more jobs. If, however, lower prices do not have a significant impact on sales the firm's revenue is likely to fall and the workers' sacrifice in taking lower pay will not be rewarded by additional jobs. Of course, if we can produce and sell more goods without any fall in selling prices, then lower wages will make it attractive to hire more workers.

Employers' requirements for labour are not entirely a function of wage costs, however. Some skills are indispensable to a business and the demand for this type of labour will be relatively inelastic. High rewards to top executives have often been justified on the grounds that top managerial talent is essential and hard to find, as well as on the grounds of their contribution to company earnings. On the other hand they have more influence than most employees on deciding their pay levels. On site skilled tradespeople are less easily replaced than labourers. Where jobs can be mechanized or done away with altogether workers are most vulnerable to losing their jobs if they push for higher wages.

In addition to the quantity of labour available, we must consider its quality, for output depends as much on the ability of the workforce as on sheer numbers. Education and training, skills, motivation and management are all important to the productivity of labour. This will be discussed further in Chapter 15.

CASE STUDIES

A.

An incentive bonus scheme: perceptions and reality

A study of piling operations undertaken to investigate the link between productivity and incentive payments began by asking management and operatives about the existing bonus scheme. Both were agreed that the effort of the workforce was a key factor in achieving high rates of production. The workers themselves were equally sure that their bonus earnings reflected the effort they put into the job.

The study examined weekly production figures and bonus payments going back several years in the firm's records and found no evidence to suggest bonus payments improved productivity. Indeed there appeared to be little if any correlation between bonus earnings and output. The main reason was inherent in the nature of the piling operation itself. The whole process revolved around heavy plant. Ground investigation drills, trenchers, augers and pile drivers were essential and progress depended on the rate at which they could be operated. The actual pile driving was the crucial element but the pace of this was dependent not on the workers' efforts, but on the state of the ground where the operation took place.

Another critical element was time lost through breakdowns. The skill of workers in getting machinery back to work was of the utmost importance but was not reflected in bonus payments. In short, output was more dependent on the nature of the activity than the degree of effort expended by the workforce.

Based on a case study recorded in IOB Site Management, *Paper No. 78, 1979.*

B.

Table 5.2 shows a breakdown of the civilian labour force by age, which reflects the changing age structure of the population at large. The labour force includes all those of working age who are available for work, whether employed or not. It excludes full-time students, housewives and others who are not in the labour market. Relatively high birth rates in the 1960s increased the numbers of 16-year-olds entering the labour force in the late 1970s. After 1981 this trend started to decline. The average age of the population and the labour force is now rising. In 1986 a quarter of the labour force was under 24 years old, by 2001 only one-sixth will be under 24 and more than a third will be over 45.

Table 5.3 shows changes in activity rates, i.e. the proportion of those in the age groups shown who are part of the labour force.

Table 5.2 Trends in UK civilian labour force: by age (millions)

	16–24	**25–44**	**over 45**	**Total**
Estimates				
1986	6.3	12.7	8.5	27.6
1990	6.0	14.0	8.9	28.9
1991	5.7	14.1	8.7	28.8
Projected				
1996	4.9	14.5	9.6	29.0
2001	4.8	14.7	10.0	29.6

Reproduced from *Social Trends*, CSO, 1993.

Table 5.3 Trends in UK civilian labour force: economic activity rate

	16–24 yrs		**All over 16 yrs**	
	Male %	**Female %**	**Male %**	**Female %**
Estimates				
1986	80.5	70.2	73.8	49.5
1990	81.7	73.2	74.3	52.8
1991	80.4	71.6	73.8	52.5
Projected				
1996	78.7	70.2	73.2	53.5
2001	78.0	70.7	72.6	54.9

Reproduced from *Social Trends*, CSO, 1993.

WORKSHOP

1. With the aid of diagrams explain the likely effect of an increase in labourers' wages on the demand for labourers.
2. If the rate of pay for crane operators went up, would you expect the effect on the demand for crane operators to respond as it did for labourers in question 1?
3. How does the analysis of marginal revenue productivity help to explain wage differentials, e.g. why a crane operator earns more than a labourer?
4. With reference to case study **A**, indicate some of the difficulties in measuring workers' productivity and using such information as a basis for payment schemes.
5. Describe and account for the trends shown by the data in case study **B**.
   Discuss the implications for the supply of labour to the construction industry.

DISCUSSION QUESTIONS

- 'A rational wage structure should reward effort not output.'
- Is it preferable to pay labour by piece rates (according to output achieved) or by day rates (according to hours worked)?.

6 Factor markets: capital

PREVIEW

- What is the link between using 'capital' to mean real assets and to mean money?
- In what sense is the price of capital the same as the rate of interest?
- Why do entrepreneurs invest in capital goods?
- How do interest rates affect investment?

The word **capital** is used in many different senses. We often use it to refer to our savings, meaning money in the bank; other times we talk of having capital tied up in property or other physical assets. In the same way investment is sometimes used to mean putting your money in a building society or bank deposit to earn interest. More properly this is **saving**. It only becomes **investment** when someone uses that money for a new asset, a piece of equipment or a building which is used in production and thereby generates earnings.

CAPITAL AS A FACTOR OF PRODUCTION

As a resource used in production, capital refers to physical assets which have been produced in order to assist the production of more goods. Tools, machinery, buildings, transport systems, stocks of components and materials all come into this broad category. Some of these goods, the **infrastructure** of the economy such as roads, schools and docks, are termed **social capital**. These assets benefit the community as whole, rather than individual citizens or businesses. Much of this social capital is provided via the state.

Business capital is the assets owned and operated by firms. It consists of **fixed capital**, e.g. buildings and plant, and **working capital**, e.g. stocks of materials, part or fully finished goods and liquid reserves (money). The latter is not making a direct contribution to production in the same way as machinery or materials, but it gives the firm access to whatever resources it needs; it is a liquid asset, providing flexibility. Capital goods must be financed from profits, by borrowing, or through selling shares. 'Raising capital' thus means getting the funds to purchase capital goods and get the business going. A firm needs financial capital to invest in the real capital assets which make up its production facilities.

THE COST OF INVESTMENT IN CAPITAL GOODS

Finance for capital investment will have to be paid for; its price is the rate of interest charged by the lender. Even if investment is financed from the firm's accumulated profits, without recourse to borrowing, there is a 'price' attached, since the profits could be loaned to others, thereby earning interest. The lost interest is the opportunity cost of investing in the firm itself. It provides a yardstick to judge whether this investment is the best way of using available funds. If the firm will make more profit from new equipment than by lending the same sum to earn interest, then the investment will go ahead.

The rate of return on capital investment is its earnings expressed as a percentage of the outlay. If a piece of equipment costs £50 000 and is expected to bring in a revenue of £7500 p.a. after operating costs have been deducted, the rate of return is given as 15%. As investment levels increase, returns on capital goods are likely to fall, because capital, like other factors, is subject to the law of diminishing marginal returns. If a civil engineering company with a fixed number of motorway contracts buys more and more earth-movers, it will eventually find it has too little space and ancillary labour to keep all the machines operating fully. Additional machines stand idle for longer spells and the rate of return starts to fall.

The earnings, or yield on the monies invested, expressed as a percentage of the last unit of capital employed is termed the **marginal efficiency of capital (MEC)**. As the firm's stock of plant gets bigger and diminishing marginal returns are experienced, MEC starts to fall. When the rate of return (MEC) drops below the rate of interest (the price of financing new investment) spending on capital goods ceases to be worthwhile. The MEC is thus the effective demand curve for investment, showing how much capital spending will be worthwhile at a given rate of interest, or price (Fig. 6.1).

The classical economists saw interest rates as a straightforward market price. Borrowers had to compensate lenders for their sacrifice of current spending power and also offer some reward for the risk that all or part of the loan might not be repaid. Interest rates depended on the demand and supply of loanable funds, with variations in rates accounted for by the differing degrees of risk and liquidity entailed in the project. Long-term loans

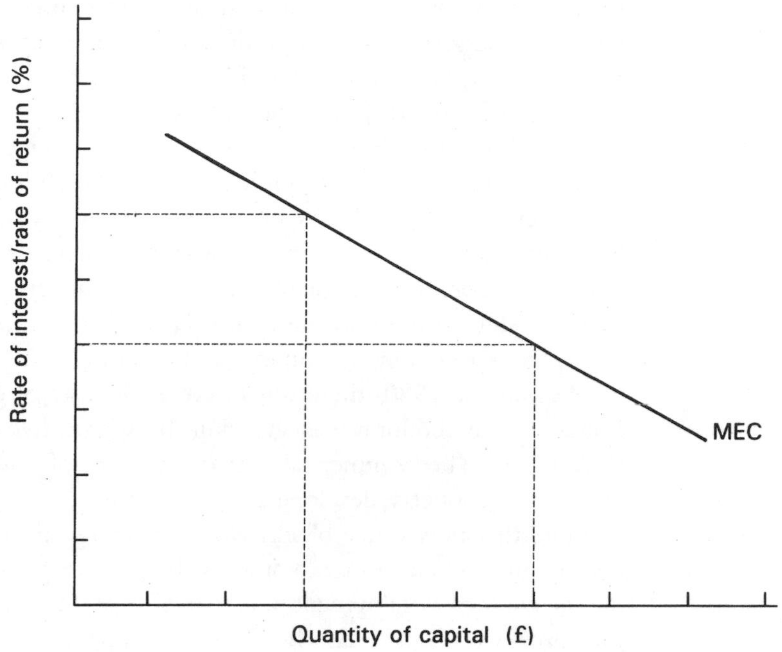

Figure 6.1
Marginal efficiency of capital.

for a project like the Channel Tunnel would earn higher interest, owing to the uncertainty of success and the length of time the money would be tied up, compared to a short-term loan for housing development.

DEMAND FOR CAPITAL

The more opportunities for profitable investment, the greater the demand for funds. As the demand for funds expands it pushes up the price of borrowing, so the general level of interest rates goes up. Opportunities for investment arise in many ways.

New technology, e.g. the development of information technology, has led to a large amount of investment in recent years. This affected the construction of commercial buildings as firms sought to incorporate IT in their space requirements. It has even had some impact on housing (see case study A, Chapter 3). The housing boom of the 1980s created opportunities for investment in building homes, especially for first-time buyers and sheltered accommodation. In the 1970s huge amounts of investment resulted from the discovery of oil and gas under the North Sea, with consequent construction of drilling rigs, pipelines, refineries, etc.

The demand for capital is derived from its use in the production of goods and services. The entrepreneur's aim is to make a profit. How much capital to employ can be tested by the same marginal analysis that was used to solve the problem of how much labour to employ in the previous chapter. The marginal revenue product of investment is compared with its marginal cost. An investment which is expected to yield 12% return will not be worth undertaking if the interest rate is 15%, but becomes profitable if rates fall to 10%. However, interest rates can change frequently and rapidly, whereas most investment projects involve much longer time horizons. A developer contemplating a scheme that will take years to complete would be foolish to embark on it if a small rise in interest rates would render the scheme uneconomic. Inevitably there is some uncertainty about any project which is still on the drawing board. Decisions to invest depend on judgement, on expectations and on experience, not just on interest rates.

Thus, in the 1980s there was a severe shortage of office space in central London. The economy was growing, businesses needed more space, rents were rising. The commercial climate was full of confidence. In such circumstances property developers were optimistic in their projections of the incomes that new office blocks would generate. By the end of the decade the mood was changing. Not only had interest rates gone up, pushing up the costs of development and reducing profit margins, but the demand for space was falling. As the economy slipped into recession firms laid off workers and cancelled plans to move into new accommodation. Quite rapidly the mood swung from optimism to pessimism. Developers downgraded their expectations and recalculated their sums with lower rents. Fewer projects looked viable in the gloomy climate of the early 1990s and the pace of development slowed drastically. As the economy began to recover and businesses regained confidence, their views of the potential returns from investments became more optimistic (see Fig. 6.2). Expectations of a higher MEC will increase the volume of capital investment.

All investment looks to the future and as the climate of opinion can change very quickly, many observers see expectations as more significant than interest rates in governing levels of investment. Even high interest rates, it is argued, will not deter investors who are optimistic about economic growth. Certainly low interest rates are not enough to make people invest in capital spending if they lack confidence that it will help their business make a profit.

Capital is often described as the most mobile of the factors of production. Clearly many physical assets – roads, bridges, buildings, large plant installations, etc. – are quite immobile. What can move easily is the financial capital needed to fund these assets. In a world with increasing freedom from restrictions on the movement of funds, the market for finance is becoming more international. Financial skills are as important as technical expertise to the success of major infrastructure projects, which often cost

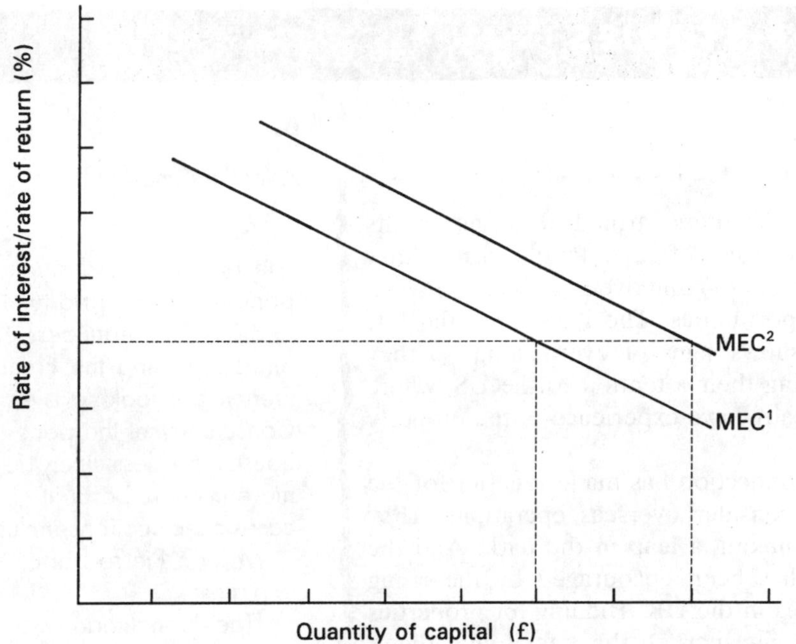

Figure 6.2
Impact of increased optimism on investment.

large sums of money. Many poorer countries which lack well-developed financial systems rely on contractors to organize funding as well as manage the construction process. Successful bidding in these international markets requires a wide range of commercial, financial and technical knowledge.

CASE STUDIES

A.

The Lee brothers, armed with the profits from the sale of the family firm, Imry Property, are looking overseas for new investment opportunities. The Lees think the UK market shows signs of overheating, so they are turning their attention to the US, where they already have experience of the property market.

That connection has made it easier for the Lees to consider overseas operations. They are not making a leap in the dark. And the process has been encouraged by the rising unit prices in the UK. Bidding for properties on the home market, the Lees found themselves second or third, and, once, 20th.

They accept that the investment market is strong, but, with special reference to central London, made the point that 'there is so much capital coming into England because it is seen as a safe bet. That's keeping the market buoyant. Whether the underlying factors are there is another question.'

Overseas the yield structure of the different property industries is different and this is what attracts the Lees. They like the ability to buy into markets where the rents are not inflated. They like an 8% yield and lower borrowing costs than currently exist in the UK. While the yields in the US tend to be higher, the downside is that leases are shorter; it is necessary to work much harder at property to make it pay.

Based on an article in Financial Times, *© F.T., 13 January 1989, reproduced with kind permission.*

B.

A 'cellars' market?

The prospect of a single European market is bringing new opportunities and new threats to British housebuilders. Beazer and Bovis are already eyeing the French market, whilst the French are looking over prospects in Britain. Consumers set the pace in the French market; houses are less likely to be built on estates, more likely to be built to order. In 4 out of 10 cases specifications include a cellar.

M Jean Pierre Babu, of Maison Bouygues, explains:

'The Frenchman wants to make optimum use of his habitable space and using it for storage is inefficient, bearing in mind the relatively low cost of providing storage space in the cellar. Cellars are simple to design and build. Windows and doors are kept to a minimum, floors are simply 80 mm or so of poured concrete and there is no need for insulation or services other than electricity ...

'This space can be used not just for traditional wine storage, but for gardening tools, general storage, as a workroom or studio, for garaging or even as a children's playroom.'

Cellars, whether room sized or the full plan area, are not normally classed as habitable space. The cost of a full size cellar on a house giving 100 sq. m of habitable area is an extra 20%. The main part of this is the excavation and additional depth to the walls. A frequent practice is to use the excavated soil to raise the ground level so the house is built on a mound. This keeps the cost even lower and has the advantage of making access to basement garages a lot easier. M Babu is confident the French will not relinquish their cellars. Will the single market bring a renaissance for cellars in Britain?

Adapted from National Builder, *October 1989, the then monthly journal of the Building Employers Confederation. Reproduced with permission.*

C.

'Large enough to serve, small enough to care.'

You have to be confident you have something special to offer to put your money into manufacturing construction equipment at a time when the industry is the trough of recession. The boss of MF Industrial, who announced a management buy-out in September, has that confidence. Richard Robson, managing director, explained why.

Firstly, a lot of painful rationalization had been achieved before the buy-out, bringing the labour force down to less than half its previous size. Saving labour costs is only half the story, the other half is investment. Manufacturing has been radically reorganized to improve the flow of production. New machine tools, paint booth and robots have been introduced to the tune of £6 m with plans to spend another £1 m on a separate parts operation to back up sales.

MF Industrial aims to reach second place in the European market for backhoe loaders by 1994. For this it will need more than an efficient factory and a good marketing slogan. Two years of development time have produced the series T Servotronic, a backhoe loader with advanced electronic control systems to improve manoeuvrability and safety. With products like this Mr Robson is ready to take on the opposition.

WORKSHOP

For questions 1–3 refer to case study **A**.

1 Why have the Lees decided to invest in US property rather than in the British property market?
Would you describe the purchase of existing property as an example of 'investment'? Explain your answer!

2 What are the 'underlying factors' referred to which affect the strength of the property market?

3 What considerations would you think relevant to choosing which overseas market offers the best investment opportunities for property companies?

For questions 4–6 refer to case study **B**.

4 What are the economic arguments for incorporating garage/storage space into a cellar rather than adding to the ground plan of the property?

5 How will material and labour costs compare for a cellar space or an integral garage/workshop built above ground? (You are not expected to provide detailed figures, but to make a broad comparison of requirements, assuming equal provision of space.)

6 Can you suggest why cellars are more commonplace in France than in Britain?

For question 7 refer to case study **C**.

7 Describe the main forms of investment undertaken by MF Industrial in case study **C**. What were their objectives? How sensitive do you think this investment would have been to interest rate changes?

DISCUSSION QUESTION

❑ British house buyers in the past have tended to view property as an investment purchase. French and German house buyers tend to regard houses as consumer goods; their homes are for living in rather than appreciating in value. What are the implications of these attitudes for the house building industry?

7 Competition and monopoly

PREVIEW

- Why do we normally think of competition as a good thing?
- Why do some firms agree not to compete with each other?
- Is the lowest price for a job the same as the best value?
- If we can't 'shop around' how can we be sure of getting value for money?
- Will prices be higher where there is a monopoly?

THE ROLE OF COMPETITIVE MARKETS

In a perfect market where competition flourishes, price quickly reflects any change in demand or supply. The price mechanism serves a dual role here. Firstly it 'rations' available supplies among would-be consumers; when there is a shortage prices will rise, making consumers economize on their purchases. Secondly higher prices channel revenues, and hence resources, towards suppliers of those goods which consumers prefer, a preference shown by their willingness to pay. In theory at least the consumer is king. The price mechanism ensures that, as long as they are willing to back their wishes with enough cash, consumers can be sure those wishes will be met. Competition makes sure that firms keep their prices as low as they can while still covering costs, including enough profit to keep the business going. In a perfect market, prices reflect consumer preferences and resources are allocated accordingly to achieve maximum utility.

Few markets, other than some world commodity markets and stock exchanges, operate at a level of efficiency that comes close to the model of perfect competition. In reality the market is often dominated by a single firm, or a small group of suppliers. These situations are termed **monopoly** (where there is only one supplier) and **oligopoly** (where there is a handful of suppliers). Here firms can exert far more power than they can in a perfect market. As there are few other suppliers they are able to create shortages by restricting their output and so command higher prices. Consequently costs of production are no longer the sole determinant of the firm's prices. The extra revenue obtained does not benefit the consumer, but creates excess profits, or goes into protecting the dominant firm's position by setting up barriers against potential competitors. The allocation of resources no longer reflects consumer preferences properly.

Supporters of the market economy place much emphasis on the benefits of consumer choice and the efficiency of resource allocation under the spur of competition. It is argued that competition keeps firms on their toes, it weeds out the inefficient and makes sure suppliers stay responsive to consumers. Monopoly allows the supplier a stranglehold on the market. Waste and inefficiency go unpunished because consumers have to accept what is offered. Otherwise they go without. These arguments have a lot to commend them and many countries, including the UK, have laws to ensure that competition is maintained.

COMPETITION NOT ALWAYS PREFERABLE

However, there are circumstances where consumer interest may be better served by a monopoly or oligopoly than by a doctrinaire insistence on competition. For instance, the provision of electricity through a single national grid has generally been thought a more efficient use of resources than having a number of firms build their own overlapping and competing distribution networks. This situation began to arise in the early days of railway building, when lines were being constructed to different specifications in different parts of the country. Of course, a single electricity authority or unified railway enterprise is not the only way to achieve an integrated network, but it may be the simplest way and thereby benefits the consumer.

Monopoly may also be preferable to competition where the market is limited and economies of scale (see Chapter 10) are potentially important. One large producer can often be more efficient than a number of smaller ones. Table 7.1 shows the volume of output for the smallest size of plant estimated to be cost efficient in a number of industries. Where this is large in relation to the size of the market monopoly may be preferable to operating a number of smaller plants in competition. Even where the efficient plant size is small competition is not guaranteed, since a single firm can run numerous production units. Where a firm uses advanced technology and relies on a secure profit base to finance expensive research programmes, too much competition could be harmful.

In a shrinking market competition may again do more harm than good. Too many firms competing for too few customers calls for rationalization. As businesses undercut each other and profits turn to losses, they may try to limit competition and reduce output to a level where normal profit can be restored. Takeovers or market sharing agreements are likely to occur,

Table 7.1 Minimum efficient plant size

Product	Estimated output of MEPS p.a.*	MEPS as % of UK produced sales
Bricks	50–60 m	0.7
Cement	2 m tonnes	11
Commercial vehicles	20–30 000 units	about 40
Plasterboard	18–20 m sq.m	17–19
Steel	4–9 m tonnes	17–37
Tractors	90 000 units	76
Turbo-generators	6000 MW	120

*Estimates vary substantially for some industries. Figures from Monopolies and Mergers Commission Green Paper, HMSO, 1987. Reproduced with the permission of the Controller of Her Majesty's Stationery Office.

sometimes sponsored by the state, in an effort to prevent wholesale collapse of the industry. British shipbuilding is just one example.

CONTROLLING MONOPOLY POWERS

UK competition policy allows the circumstances of each case to be considered before deciding what action to take. The law is predicated on the assumption that monopoly is undesirable, but firms have the right to present arguments showing why it is in the public interest that their particular monopoly should be allowed. The Director General of Fair Trading is responsible for competition policy and acts as a watchdog for consumer interests generally.

Investigations of monopolies, including state-owned enterprises and merger proposals which would result in monopolies, are undertaken by the Monopolies and Mergers Commission (MMC). This body can look only at cases referred to it by the Director General or the Secretary of State for Trade, and its reports are advisory rather than binding. Monopoly is defined here as a market share of 25% or as control of assets in excess of £30 m in the hands of a single firm or more than one firm acting jointly.

The central issue is whether or not a firm has acted in a way that is detrimental to the public interest. The term 'public interest' was not very clearly defined in the original legislation and recent laws have emphasized 'competitiveness' as the criterion to look for – is there sufficient competition to make firms responsive to consumers and to keep them from becoming complacent?

Decisive evidence is less important than sound judgement, as the question is a matter of balancing what is now the case with what might happen if the monopoly/merger were broken up. Has the firm acted effectively in the past? Has it responded to consumer complaints? Has it brought new products to the market? Has it invested in new production technology? Or has it raised profit margins and spent heavily on advertising to keep up sales? Will a merger lead to increased technical cooperation and economies of scale, giving consumers better, cheaper products, or will it mean less consumer choice and higher prices?

Collusion can be just as harmful as outright monopoly. Any agreement made by firms to limit the competition between them has to be registered. It can then be scrutinized by the Restrictive Practices Court, which dates from 1956. Such agreements include fixing prices, deciding on areas or customers to be served, or agreeing conditions of sale. The grounds which a firm can use to defend its agreements are closely specified and it must also show that benefits to the public outweigh any disadvantages from the reduced level of competition. In fact most agreements put on the register have been abandoned without ever being tested by the court. Those which have been examined have been rejected in the majority of cases. The court's decisions are legally binding.

The main 'gateways' through which a firm may seek approval of its agreements are:

1 It protects the public from injury, e.g. from faulty installation or misuse of goods.
2 It provides 'specific and substantial' benefits which must be described.
3 It is needed to counteract restrictive practices operated by others outside the court's jurisdiction.
4 It counterbalances a monopoly on the part of suppliers or customers of the parties concerned.
5 Its removal would bring about serious and persistent unemployment.
6 Its removal would substantially reduce export earnings.
7 It upholds another agreement already recognized by the court as being in the public interest.
8 It does not materially reduce competition.

The sale of goods to the public is covered by similar legislation. Suppliers cannot fix the retail price of their goods unless they can show it is in the public interest to do so. Again the possible 'gateways' are laid down. They are concerned with maintaining the variety and quality of goods, and ensuring an efficient distribution system and an adequate level of service to guard against misuse and danger to the public.

EUROPEAN COMPETITION POLICY

New rules came into force in September 1990 to regulate competition throughout the then European Community (now known as the European

Union). These make the European Commission responsible for vetting larger mergers, where the combined turnover exceeds ECU 5 bn (approx. £3.51 bn), with a European dimension. The EU's main concern is to judge whether the outcome is likely to impede 'effective competition within the market'. On the surface this appears very similar to the UK position, but where UK policy emphasizes competitive structures, EU policy places more stress on competitive behaviour, which can include such things as 'the development of technical and economic progress'.

Under article 86, dealing with abuse of monopoly powers, no minimum market share is laid down. By focusing on the abuse of monopoly rather than its existence EU law lays the emphasis on the actual effects of what firms do. Decisions can be appealed within the European courts but are not subject to political review. Smaller mergers and those which do not involve European interests continue to be regulated by member states. Article 85 is concerned with restrictive practices of the same sort that UK legislation seeks to prevent. Exemptions are possible for agreements which are considered beneficial but the penalties for operating agreements which have been found contrary to the public interest are substantial – fines can be imposed of up to 10% of a firm's turnover.

For the largest and the smallest mergers there is no problem deciding which authority is responsible. For some intermediate cases it may be less clear cut. Firms which fail to notify the EU of proposals which fall within their jurisdiction are liable to severe penalties. Where EU rules do apply they take precedence over national judgements, although appeal can be taken to the European Court of Justice. The aim is to streamline procedures, so that deals can be vetted by a single authority, operating to a strict timetable in the case of the EU. Whether the demarcation between Brussels and the national authorities is sufficiently clear for firms to avoid applying for clearance to both remains to be seen. Even as Brussels prepared to take up its new role, in London three major cases were referred to the UK's MMC in what was widely seen as an attempt to pre-empt the exercise of the new Community powers. However, the proposed merger between Tarmac and Streetley in 1992, which clearly came under EU regulations, was sent back to the UK because although it created a dominant share in the domestic market, it did not materially affect competition in the Union – further evidence of Brussel's emphasis on the impact, rather than the form, of the cases before them.

The effectiveness of EU policy will depend on how fairly and consistently it is applied. The multinational dimension implies more cooperation between the EU and member states than always evident in the past. So far EU policy has avoided the vagueness of public interest arguments, but has concentrated on issues of how far competition is affected by the monopoly/merger in question. The legislation and the powers to ban collusive behaviour are stronger than UK domestic roles, but they are also more closely defined and less open to the exercise of discretion.

CASE STUDIES

A.

The Monopolies Commission investigation of the London Brick Company found no evidence of the company having exploited its dominant position in the market for fletton bricks. Profits were not found to be excessive, averaging 23% return on capital over the past twenty years. During this time profits had never topped 30% and, given the fluctuations which were a hallmark of the market for bricks, the firm's record was surprisingly consistent.

However, the company's policy on transport charges did cause the Commission some concern. The policy originated when fletton bricks were first introduced to the market. In order to make them more widely acceptable the company had reduced delivery charges on long hauls and compensated by raising prices for local deliveries. This structure of delivery charges had been maintained even though the fletton was now a well known product throughout the country. The company felt the practice was justifiable because it ensured prices did not differ greatly from one region to another. Maintaining the volume of sales helped to secure economies of scale. Despite this argument the Commission concluded the policy was against public interest.

Based on a report issued by the London Brick Company to its shareholders.

B.

Collusion has been a commonplace and persistent feature of the construction industry. The Restrictive Practices Court investigated 67 agreements between contractors in the years 1969–72. Injunctions were issued against 99 firms as a result. Twenty years later the problems remained [see case study **C**]. All of the 67 agreements involved market sharing arrangements of the following types:

- Cover prices, e.g. electrical installations at Smithfield market. Contractors took 'cover' prices from one firm. By pricing their own tenders above the cover price the others failed to offer competition.
- Tabling prices, e.g. mechanical services, HMS *Dryad*. The firms invited to tender agreed to meet and reveal their bids. The lowest estimates would be ruled out and all firms agreed to put in bids above the one selected to go forward to the client.
- Compensation payments, e.g. London Airport contract. Five firms invited to tender agreed that the winner would pay £5000 towards the cost of each loser.

C.

Unlawful cartels uncovered by the Office (of Fair Trading) in recent years include price fixing or market sharing agreements for: reinforcing steel bars, black top road surfacing materials and ready-mixed concrete.

Sir Gordon Borrie, Director General of Fair Trading, has persistently criticized the construction industry, which he says has the worst record of any sector for such practices.

The Office of Fair Trading, in an attempt to deter local managers from operating unlawful cartels, has started to take individuals to court as well as companies. The Restrictive Practices Court has the power to impose prison sentences of up to two years.

However, Sir Gordon has been disappointed by the reluctance of customers of the construction industry to seek damages through civil courts against cartel operators. He says, 'local authorities have a legal requirement to pursue value for money on behalf of taxpayers but I am aware of only one case where a council started legal proceedings against members of a cartel and that was settled out of court.'

The Office will shortly issue a guidance note to local authorities informing them how to look for clues such as unusual pricing patterns to spot cartels, particularly in construction. Sir Gordon believes the lack of an effective response by customers is one reason why cartels persist in the industry.

Extract from Financial Times, © F.T., *25 July 1990, reproduced with kind permission.*

D.

Recession in the construction industry has severely affected the suppliers of construction equipment, an area which has repeatedly suffered from problems of overcapacity. Even as the industry's shop window displayed its wares at the 1992 Bauma exhibition in Munich, one observer estimated that dealers around the world had their yards packed with sufficient hydraulic excavators to supply the construction industry for the next six months. Competition is so fierce that buyers commented, 'You can virtually name your price'.

Caterpillar, a world leader in construction plant, has recorded its first loss, a loss of over $400 m, since 1984. The industry is no stranger to rapid transitions from boom to bust trading conditions. It remains fragmented; too many firms scrabble for business, competing instead of cooperating, to maintain their stake in the big markets, Japan, Europe and North America. The result is an ailing industry suffering from up to one fifth excess capacity.

WORKSHOP

1 How would you define 'the public interest'?

2 What evidence did the Monopolies Commission cite in support of their view that the London Brick Company had not exploited their position unfairly?
Can you suggest other sorts of information which would provide evidence to support or refute allegations that a company acted against the public interest?

3*How might companies involved in the collusive tendering practices, described in case study **B**, defend their actions?
Outline the arguments against each of the practices.
Explain whether you consider such practices are against the public interest or not.

4 What reasons can you suggest for the high incidence of restrictive practices in the construction industry?

5 The excess supply of construction equipment (see case study **D**) created a buyer's market with rock-bottom low prices and profits.
Is this in the public interest?
How would customers and suppliers benefit from less competition and more cooperation between manufacturers?

DISCUSSION QUESTION

'The construction industry is highly competitive.' 'Collusion in the construction world is endemic.'

❑ To what extent do you agree with either, or both, of these statements?

8 Theory of the firm

PREVIEW

- Why do some people choose to start their own businesses?
- Are the aims of a business the same whether it is a small, family firm or a large company?
- How consistently do managers pursue business objectives?
- Whom do firms seek to please?

FIRMS AND THEIR OBJECTIVES

Classical theory starts with the assumption that the aim of the firm is to maximize profits. Perhaps this seems obvious! Profit is certainly a condition of survival in a market economy, so it must be one objective, perhaps the most important one, but not necessarily the only one. Many firms seek to achieve a **satisfactory** profit alongside other targets. Targets are set by those in control of the business. Traditional theory assumes owners and decision-makers are the same people, but this is not always so.

A firm's behaviour is open to many influences, including its size, its history and traditions, its organizational structure, the type of product or service it provides, the market structure in which it operates and the social and political ethos of the day.

The model of perfect competition is based on profit maximization. In a highly competitive market, with a single type of product and large numbers of small firms, each one must be as efficient as possible to survive. There is no room for objectives other than profit. This model becomes less realistic as we relax the assumptions of perfect competition. The growth of big business in the twentieth century has altered the balance of power between firms and customers. Firms that can differentiate their products and dominate markets have more independence in setting their prices and objectives. We look at alternative theories of the firm which take this into account here; the traditional theory is explained in the next chapter.

With small firms the person in charge is likely to be the owner, so the link between management and reward is very immediate. In this situation we might expect the profit motive to retain its priority. But investigations show people have a variety of reasons for running their own businesses. Many simply wish to be in control of what they do. They are individuals who dislike the bureaucratic environment of larger organizations. Others go into business because they cannot get new ideas or new products developed by established firms. Some want to maintain traditional skills and products, which larger firms have abandoned. Many see self-employment as a more secure option than depending on an employer for work. Others are pushed into going it alone through redundancy. Some enjoy the challenge of creating a business from scratch. A few cherish the hope of making their fortune. Clearly making a profit is necessary to achieving any of these aims, but only the last requires profit maximization!

SATISFYING DIFFERENT INTERESTS

In large businesses, which dominate many industries today, it is less easy to identify the source of their motivation. The majority of these firms are public companies, owned by shareholders but run by salaried managers responsible to the board of directors. Who is in control, who sets the goals? In major companies, with shareholders running into thousands, effective control by so many individuals is difficult if not impossible. However, most shareholders are content to take no part in the firm's affairs as long as they receive satisfactory dividends. The directors are elected by the shareholders to oversee the company's affairs. Directors may be chosen for their outside connections and status to add prestige to the board. Provided the business does not run into any crises, they are unlikely to intervene. This leaves the chief executive and management, who are paid to run the firm's operations, to determine its goals also. What motivates the managers?

So long as they achieve sufficient profits to keep the shareholders happy, managers are likely to focus on aspects of the business which more obviously reflect their own skills and bring them success in their careers. The measure of success varies according to the type of business. Retailers like to see growth in sales figures, an ever expanding turnover. High technology firms emphasize research and innovation. Contractors may seek to win contracts in high prestige developments, from digging the Channel Tunnel to rebuilding the National Gallery.

Growth of the business, whether measured by turnover, capital value, number of employees or market share, is a common aim. The bigger the business, the more power and status for those running it. Family firms – and there are many family firms in the construction industry – are frequently less concerned with growth. They may fear that, beyond a certain size, they risk losing control of the business. They are often more concerned with continuity than expansion, wanting to hand on the business to the next generation. A solid reputation and a secure customer base are important in their search for survival. This type of family business is more often found among the smaller and medium sized companies than the bigger public companies.

Increased professionalism among managers has aggravated fears that the divorce between ownership and control is becoming entrenched in larger companies, to the detriment of shareholders. Management expertise is not easily challenged by non-specialist shareholders. Against this it can be argued that the growth of institutional shareholders, such as unit trusts and pension funds, has redressed the balance in recent years. These funds are also managed by professionals, who use their powers as shareholders more actively than most private investors.

This raises an interesting question as to whether groups of shareholders may have different objectives. The dominance of professional investors has its own dangers. Fund managers, who are judged by the growth of their funds, may prefer the firms they invest in to pay large dividends rather than reinvest profits in long-term development. Short-termism is a complaint levelled against British industry, often in contrast to attitudes found in Europe. It is particularly damaging to industries such as construction that tend to suffer cyclical swings in demand and may find it difficult to keep up growth in dividends. Increasingly small shareholders are forming pressure groups to express their views e.g. to press the company to implement policies on environmental, safety or other issues.

The divorce between ownership and control is not the only conflict of interest in a company. Indeed, top managers are also shareholders in most cases, so differing aims may be more apparent between higher and middle management levels than between top managers and shareholders. A more common problem lies in the divisional structure of many large firms. A contractor with a plant division, which off-hires plant to other companies, needs to be clear about the aims of that division. When the plant manager off-hires equipment outside the firm his profits improve but this can cause bottlenecks elsewhere in the company to the detriment of its overall performance.

Different interest groups within the company have their own priorities, but even if they agree that profit is the overriding objective for all of them, problems remain. How is profit measured? Which policies are going to be the most successful? Can expenditure on a specialized item of plant, or on research, or sponsorship of a local sports team, be justified by appealing to the profit and loss account? Would a different policy have generated more profit?

The cowboy builder overcharges and moves on, hoping to make a quick buck before his reputation catches up with him. Family firms, whose owners intend to hand on the business to their children, aim to establish a reputation for good value which will secure their future. Is profit to be measured by the month, by the year, or over a longer period? Decisions are made in an uncertain world with imperfect knowledge; this makes 'profit maximization' a bit like shooting in the dark at a moving target.

Small wonder that many firms are better seen as **satisficing** rather than maximizing profit. If profits are enough to satisfy the shareholders' needs for dividends, or the owners' need for income, firms will be able to pursue alternative targets. Chief among these are growth in sales revenue. Sales figures have the advantage that they are more easily measured than profit. They provide a quick, easily understood indication of the firm's progress. Market share is another indicator, one which stresses performance relative to competitors. In addition, as suggested above, managers are likely to pursue policies which enhance their own positions, within and outside the organization. The term **managerial utility** has been coined to describe this state of affairs. Unchecked it can result in lavish expenditure on company cars, private health care schemes, foreign travel, personal assistants and a variety of perks which boost prestige and raise the effective level of pay for executives and senior managers above the acknowledged rate.

MERGERS AND TAKEOVERS

Whichever objective takes priority – profits, turnover, market share, capital value, investment – if a firm aims to maximize the figures it is pursuing an expansionary path. Growth is an implicit goal for a great many entrepreneurs. Most expansion takes place internally, with the firm using its profits to promote sales and improve its production facilities. The alternative route, actively pursued by some companies, is growth via mergers or takeovers.

Mergers involve firms getting together by mutual agreement. Pooling resources, particularly if the firms' activities complement each other, may be seen as creating a stronger business, with possible economies of scale (see Chapter 10). Takeovers are the purchase of one firm's shares by another, sometimes against the wishes of the directors. Again there may be economies of scale, but there are many other possible motives. Among the more controversial are asset-stripping (where the victim's business can be broken up and its assets sold off for more than the purchase price of the business), or a simple desire to reduce the competition. A wave of takeover activity in the early 1970s, involving a number of smaller housebuilding firms, was characterized by the desire to acquire their assets, in particular their land banks. The motivation here was to exploit these through development rather than sell them off.

Three types of merger can be distinguished. Horizontal mergers are between firms in the same type of business. The proposed merger between Steetley and Redland in 1991, both brick manufacturers, was of this type. The motive on this occasion was not growth but rationalization in the face of a declining market. Vertical mergers involve firms who are potential customers/suppliers to each other, e.g. a brick manufacturer coming together with a house building company. Tarmac's acquisition of firms engaged in the extraction of sand and gravels during the 1980s comes into this category. Lateral mergers will result in conglomerates, such as Hanson or Trafalgar House, firms which are composed of unrelated businesses. Trafalgar House owns construction, shipping and other interests.

Each type of merger has its own rationale. Mergers seem to be more frequent during periods when the economy is booming. These mergers are usually aggressive. They are seen as a quick route to growth by acquiring the production plant and customers of other firms. Increased market share, and with it greater monopoly power, is an obvious result of horizontal mergers. Vertical link-ups can also strengthen market position. In this case the firm may be seeking to control its sources of supply (or its outlets) for greater security or to exclude its competitors. Conglomerates can use diversification to reduce their exposure to risks, especially if some of their businesses are very vulnerable to trade cycles. Not all mergers are the result of a well thought-out business strategy – some appear purely opportunist, a chance to use surplus cash to show the world that the company is a force to be reckoned with!

Behavioural theories of the firm recognize that it is rare for objectives to be simple. A variety of interests, some of them incompatible, are at work in most business situations. Management means reconciling these conflicts, potential or actual. Planned objectives sometimes get pushed aside as events outside the firm's control take priority. The only overriding aim is the survival of the business. Profit is a necessary condition of this aim, but profit, growth and all other objectives only have to be pursued to the level which ensures the firm will continue to function. The idea of maximizing any single measure of success is replaced by satisficing – achieving an adequate performance across a spectrum of objectives.

CASE STUDIES

A.

At the beginning of the year Aveling Barford stepped up its attack on the articulated dumptruck market with a controversial marketing plan. The Grantham based company is offering discounts of £10,000 on AB artics to firms running competitors' trucks that are prepared to return them under a buy-back option. The tactics have drawn much criticism from the market place ... 'our primary mission is to build and sell rigid and articulated trucks', says (managing director) Wordsworth.

Reproduced with permission from Construction Weekly, *21 February 1990.*

B.

'Yes there's a rates war on. Let me make one thing clear, we didn't instigate it, but we're fighting back ... Hire firms have won contracts at 60% below normal market rates. We just don't know how they're doing it.' Speaking is David Wilson, managing director of Charles Wilson Plant Hire ... It's not only the rates war that concerns Wilson. He claims that introduction of free deliveries is tantamount to the same thing. 'As far back as I can remember deliveries have always been charged. Now free deliveries are common.'

The situation became so competitive he had to start offering the same service on small tools. Offering free delivery has also necessitated extra investment. The company purchased six extra vans at £5,000 each. 'Not a high capital cost,' says Wilson, 'but substantial running costs.'

Reproduced with permission from Construction Weekly, *21 February 1990.*

C.

In 1985 directors of the Aberdeen Construction Group, a holding company, announced a reorganization that would bring together four civil engineering and building subsidiaries to form a new, multi-skilled contracting company under the title Hall & Tawse Construction Ltd.

The four subsidiaries, which were wholly owned by ACG, but operated independently up to then, consisted of one civil engineering business and three builders, two based in Scotland, and the third, Reema, in Southampton. Together they represented 70% of the group's turnover and over half of its pre-tax profits. With a combined turnover about £80 m at the outset, the directors aimed to expand to £100 m within a couple of years.

Optimism about the growth prospects of the construction market was based on the fact that, as the chairman of ACG put it, 'so much cash needs to be spent on essential projects.' The newly integrated company was expected to benefit because 'Hall & Tawse has the means for many more larger projects than its predecessor companies.'

Bill Hendry, appointed to head the new company, was equally confident. Their intention was to target a wide spread of jobs, not just major contracts but minor works too. The new organization would have a wide range of skills to tackle any sort of work. With offices already in Scotland and Southampton, they planned to set up other regional offices and extend their geographical spread. Constituent companies already handled contracts of £20–30 m, but expected the new company to be in line for much bigger jobs, including major contracts for government departments. Mr Hendry stressed the new firm's credentials: 'The company will be directed by a highly skilled management team with proven success within individual operating companies.'

Adapted with permission from
Contract Journal, *14 November 1990.*

WORKSHOP

1 What do the extracts in case studies **A** and **B** reveal about the aims of the firms concerned?

2 Discuss the reasons for merging the four subsidiaries of ACG to form Hall & Tawse Ltd. Can you suggest any difficulties that might arise from the merger?

3 In an address given to the Joint Building Group in 1968, on the 'Ethics and Conduct of Designers and Constructors', Sir Maurice Laing spoke of the wide ranging duties of a large contracting company. He identified duties toward shareholders, employees, clients, subcontractors and suppliers, and the community.

- ***a*** What do you see as the responsibilities of the contractor to each of these groups?
- ***b*** Is there any conflict of interest between them?
- ***c*** Would you expect the priorities of firms in the industry in the 1990s to be any different from those set out by Sir Maurice in the 1960s?

4 Refer back to case study **A** in Chapter 7 on the London Brick Company. To what extent was their pricing policy consistent with maximizing profits? What other objectives might this policy be intended to achieve?

5* 'The aim of the firm is to maximize profits.' How satisfactory is this statement as a description applied to construction firms today?

DISCUSSION QUESTIONS

- ❑ Does your firm have a mission statement (see case study **A**) which expresses its fundamental business objectives? If so, how far is it consistent with the idea of maximizing profits? Does the approach to work that you have experienced within the company reflects its mission?
- ❑ If you were to consider setting up your own business, what would your main objectives be? How would you try to implement them?

9 Profit maximization

PREVIEW

- Can a firm in perfect competition choose how much to produce and what price to charge?
- Is a monopoly free to decide its output and prices?
- Is it possible to have a monopoly and at the same time to be in competition with others?
- Must all costs be covered in the short run?
- Should the costs of building the Channel Tunnel be reflected in the fares charged for using it?

Traditional theory of the firm is based on profit maximization. Profit is defined simply as the difference between revenue and costs. Revenue depends on both the volume of sales and the unit price. Since prices are affected by the level of competition, we must analyse how the firm behaves in different market situations.

COSTS AND OUTPUT

Costs are not as straightforward as they might appear. What is the cost of travelling from Southampton to Birmingham by car? It depends on what counts as a cost for the purpose of that trip. Obviously fuel costs count and perhaps an allowance for wear and tear, but do you include a share of the road tax and insurance? What about the cost of buying the car, should a part of that be included? Is this a single journey or a regular trip? It may not be possible even to identify the relevant costs without more information. The first step is to break down costs into different groups.

Some costs are **fixed** in the short run. These costs, such as insurance and depreciation, remain at a constant level regardless of changes in output. They are also called **overheads** or prime costs. Other costs are **variable**, or direct costs, so called because they are directly linked to output, e.g. labour and materials. Variable costs increase with production, but not necessarily in direct proportion to it.

To determine the point of maximum profit we focus on the **margin** because this is where adjustments take place. The question for the firm to decide is whether to increase output a little, or whether to cut back a little – how will overall profits be affected? The marginal unit of output is always the one at the point of decision. Thus if output is stepped up from nine units to ten, the marginal unit ceases to be the ninth and becomes the tenth. The extra cost incurred by producing that last unit is the **marginal cost** (MC) and the extra revenue generated by its sale is the **marginal revenue** (MR). Profits are maximized at the point where marginal cost just equals marginal revenue (MC = MR). Why?

Producing an extra unit raises costs, but it also increases revenue. If costs go up by more than is added to the revenue (i.e. MC > MR) the extra unit is not worthwhile, but if MR exceeds MC total profits will be increased by expanding output. Profit is maximized when MC = MR, the marginal unit neither increasing nor decreasing overall profit. Of course this assumes the firm is in profit, i.e. covering its total costs, so the equilibrium (MC = MR) must be equal to or above the average total costs at that point. The figures in Table 9.1 illustrate the relationship of costs and output.

Column (a) is the quantity of units being produced. Column (b) gives the fixed costs, in this case 60. This stays the same whether output is 1 or 12 units and even remains payable when output is zero. Variable costs are shown in column (c). The rest of the figures can all be calculated from the information already given. Total costs are (b) + (c). Average total costs are (d) ÷ (a); this gives the cost per unit and can be compared with the selling

Table 9.1
Schedule of costs

(a) Output (units)	(b) Fixed costs £	(c) Variable costs £	(d) Total costs £	(e) Average total costs £	(f) Average variable costs £	(g) Marginal costs £
0	60.00	—	60.00	—	—	—
1	60.00	10.00	70.00	70.00	10.00	10.00
2	60.00	17.00	77.00	38.50	8.50	7.00
3	60.00	27.00	87.00	29.00	9.00	10.00
4	60.00	40.00	100.00	25.00	10.00	13.00
5	60.00	56.00	116.00	23.20	11.20	16.00
6	60.00	75.00	135.00	22.50	12.50	19.00
7	60.00	98.00	158.00	22.57	14.00	23.00
8	60.00	125.00	185.00	23.13	15.62	27.00
9	60.00	156.00	216.00	24.00	17.33	31.00
10	60.00	191.00	251.00	25.10	19.10	35.00
11	60.00	230.00	290.00	26.40	20.90	39.00
12	60.00	273.00	333.00	27.80	22.80	43.00

price, or revenue per unit, to check whether the firm is operating at a profit. Average variable costs, (c) ÷ (a), will be needed if the firm is not covering costs and trying to minimize losses. Finally column (g) shows the marginal cost, the addition to total costs (d) of each successive unit. The marginal cost of the tenth unit is the difference in total cost between output at 10 and output minus the tenth unit, i.e. at 9 units.

When the average total cost figures are plotted on a graph they produce a U-shaped curve (Fig. 9.1). This is not surprising if the pattern of fixed costs and marginal costs is considered. Initially fixed costs are concentrated on a small output, but as output is increased the fixed costs are spread over a greater volume of production and cost per unit falls. As the firm continues to expand output, marginal costs rise. (This is due to diminishing marginal returns – see Chapter 4). These rise rapidly as full capacity is reached, pushing up total cost per unit. The bottom of the curve, where unit costs are at their lowest, is the most efficient level of output, termed the **optimum**.

The optimum is the most efficient point of production but not necessarily the most profitable. Profits depend on revenue as well as costs, so the cost curve by itself is not enough to determine the best profit position. Although costs are not affected by the degree of competition, revenues are because they depend on what price the firm can charge.

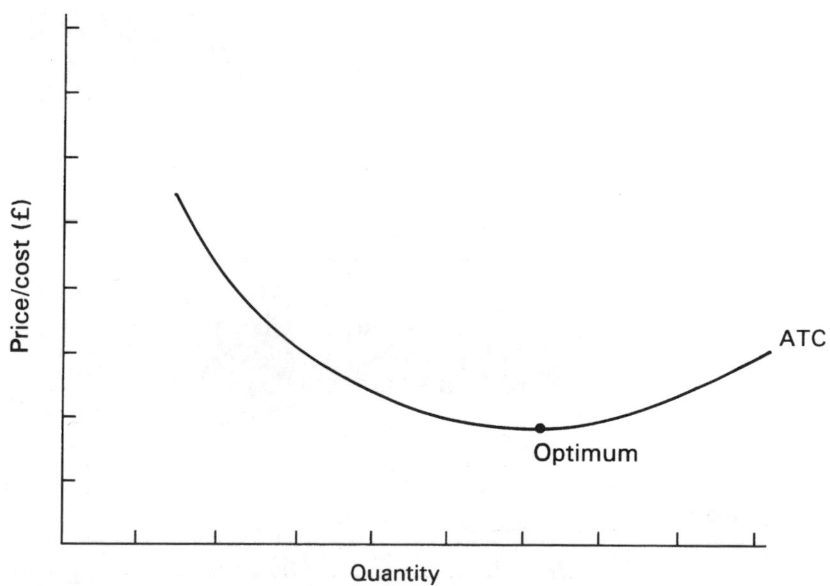

Figure 9.1 Average total cost curve.

PERFECT COMPETITION

In a perfect market the market as a whole sets prices. Each firm is a **price-taker**, unable to vary the market price. There are too many competitors to try to raise prices – customers would simply go elsewhere. Undercutting is pointless, since the firm can sell what it wants at the market price. In the long run competition keeps that price to a minimum, just enough to cover costs, including a **normal profit.** Normal profit is just enough to keep production going. It can therefore be regarded as part of the total costs and is included in ATC.

The market price is the average revenue (earnings per sale). Because the firm has no need to drop its price in order to increase sales, marginal revenue also equals price, each extra unit adding the same sum to total revenue. The two coincide, shown diagrammatically by a horizontal line which is the demand curve for the firm. The price is already set by the market – the firm has only to decide what quantity to produce. If the market price is currently above the firm's optimum the picture would look like Fig. 9.2.

The firm will maximize profits (MC = MR) at output q. Total cost per unit at this point is less than the price (revenue per unit), so the firm is making more than a normal profit. This will encourage more firms to enter the market. As competition increases, prices fall and the firm moves towards the optimum. When prices just cover costs (which includes normal profit) and firms are operating at the optimum, there is no longer any

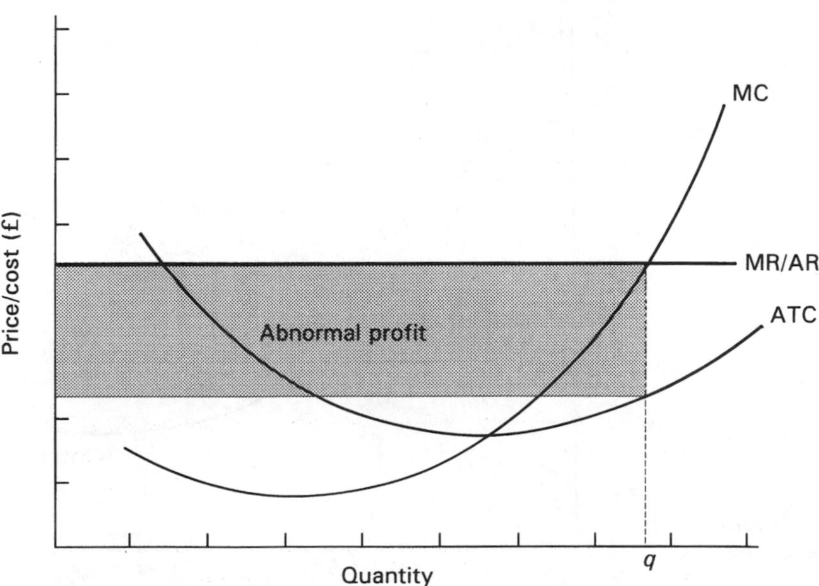

Figure 9.2
Perfect competition (short run).

incentive for more firms to enter the market. Supply and price stabilize (Fig. 9.3).

Should demand drop the price may fall below the optimum and the firm is no longer in profit. In the long run it cannot continue operating at a loss and must leave the industry. Hence the industry will contract and supplies fall until market equilibrium is restored at a lower level of activity.

Of course firms do not all decide to leave the industry immediately the price goes down. If a price fall is thought to be temporary, or if the firm can restructure its business so as to reduce costs, it can survive a period of losses. In this situation its short-run goal is to minimize losses rather than maximize profits. Once again this is achieved where MC = MR, although the firm will be operating below the ATC curve (Fig. 9.4). If the revenue is too little even to cover its variable costs, the firm's best policy is to cut out variable costs by temporarily ceasing production.

Until the firm itself ceases to exist it still has to pay fixed costs, such as rates and insurance. These are not related to production so should not influence the decision to stop producing. Like **sunk costs** they are bygones, no longer relevant to current decisions. The millions already spent on excavating the Channel Tunnel would not justify its actual use if prices could not cover operating costs. Money already committed cannot be recalled and should not influence decisions on future production.

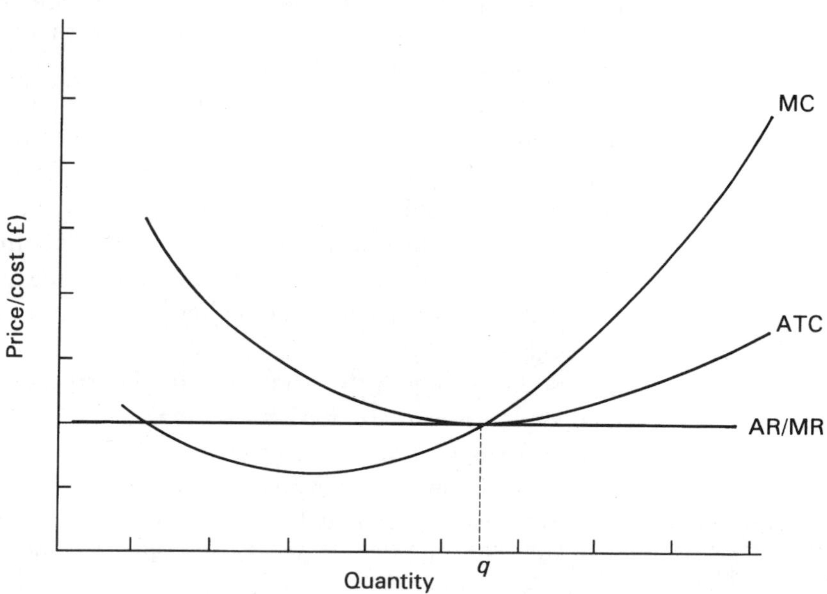

Figure 9.3
Perfect competition (long run).

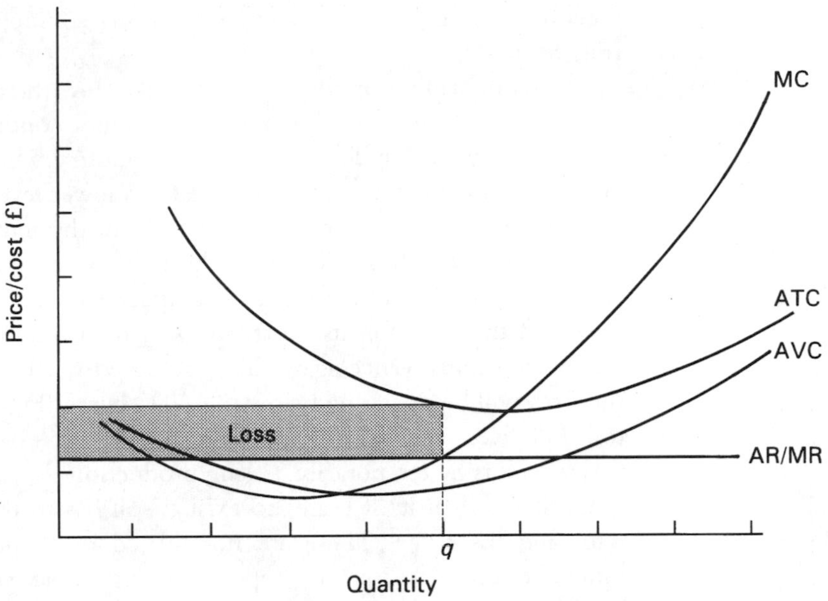

Figure 9.4
Minimizing losses.

MONOPOLY

The same rule of profit maximization, MC = MR, applies to monopolies. The difference is in the relationship between marginal revenue and price.

A monopolist controls the entire market supply and can exert some influence on prices. Monopolists are **price-makers**, but they cannot fix prices and quantities entirely to suit themselves. If they raise prices consumers will buy smaller quantities; if they wish to sell more, they must lower prices.

Table 9.2 shows how quantities sold change with price. Because the firm has to drop prices to increase sales, marginal revenue is no longer the same as price. At £27.50 four units are sold, bringing in £110. At £25 sales rise to five units, totalling £125, so the marginal revenue achieved by the extra sale is £15. The marginal revenue curve lies below the demand (or average revenue) curve. As a result the profit maximizing position, MR = MC, occurs at a smaller quantity and higher price than it would in a competitive market (see Fig. 9.5).

By restricting output, the firm can earn abnormal profits. Since there is no competition these are long-term, not temporary, profits and may be referred to as a form of rent, or surplus. Clearly the ability to earn this surplus depends on the ability to exclude competition. Entry barriers vary from informal agreements between firms not to poach each other's business, to legally enforceable privileges, such as patents. In some cases barriers

Table 9.2
Schedule of sales revenue

Quantity (units)	Price £	Total revenue £	Marginal revenue £
1	35.0	35	—
2	32.50	65	30
3	30.0	90	25
4	27.50	110	20
5	25.0	125	15
6	22.50	135	10
7	20.0	140	5
8	17.50	140	0
9	15.0	135	–5
10	12.50	125	–10
11	10.0	110	–15
12	7.5	90	–20

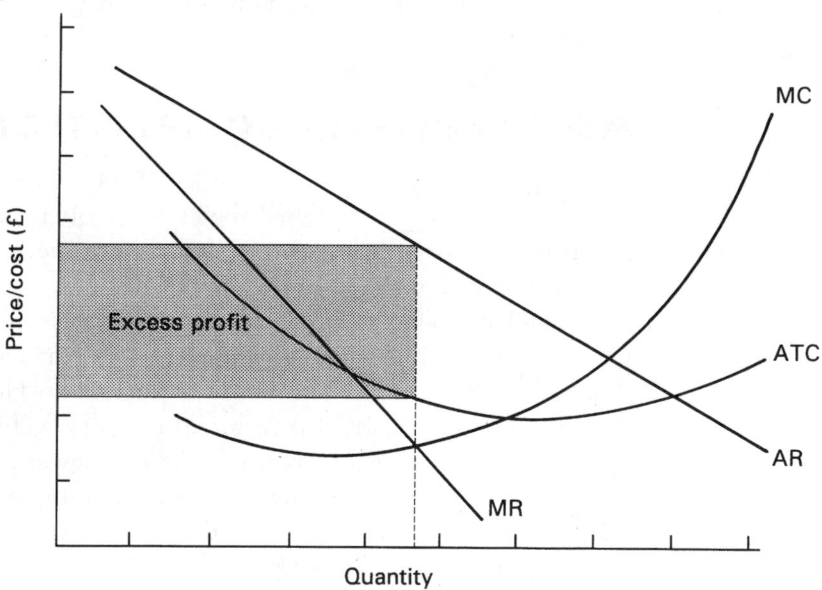

Figure 9.5
Monopoly.

are simply the high costs facing new entrants to the industry.

This analysis demonstrates how monopoly fails to serve the public interest but it should be recognized that it is possible in some circumstances to combine excess profits for a monopolist with lower prices for the consumer than could be achieved by competition. This is because a single, large firm may be able to supply the market more efficiently than many smaller ones (see Chapter 10).

DISCRIMINATING MONOPOLY

So far we have assumed that all purchasers pay the same price for the goods. If the market can be divided into separate sectors, each charged a different price, the monopolist can further boost profits and output. For this to be possible consumers must be unable to resell the goods at the lower prices. This is more often the case with services rather than with tangible goods. Transport is a case in point. Railways sell tickets at different prices for journeys at different times of day. Plant hire companies may offer cheaper rates on some tools for weekends and holidays to encourage DIY trade. The aim is to attract as many customers as possible and maximize revenues by charging each group according to their ability and willingness to pay. The variations in price are based on the realization that the elasticity of demand differs between the various consumer groups. A lower price for DIY weekenders will bring in extra business from price-conscious consumers, while a higher charge for weekday builders will maximize revenues from professionals who are earning money by their use of equipment.

MONOPOLISTIC COMPETITION

Few actual markets are quite like either of the two extremes, perfect competition and monopoly, analysed above. More often elements of both occur in what can be described as **monopolistic competition.** This is typical of markets in which branded goods compete.

A branded product is in one sense a monopoly – there is only one supplier of that brand. If 'Holdfast' is the registered trademark no one else can supply 'Holdfast' tools. The demand curve for 'Holdfast' is a normal, downward sloping curve. If their prices rise they will not lose all their sales as some customers remain loyal to 'Holdfast'. Lower prices will attract extra customers away from other brands. The relationship between marginal cost, marginal revenue and price is like that of a true monopoly. But, unlike a true monopoly, they face competition from many close substitutes, tools produced by other manufacturers.

The competition means that 'Holdfast' cannot hang on to the excess profits enjoyed by a monopolist in the long run. A high level of profit will attract other manufacturers. As the market expands prices will fall until profits are reduced to normal levels. Because the demand curve for 'Holdfast' still slopes downward (Fig. 9.6), this long-term equilibrium cannot occur at the bottom of the cost curve, so prices remain somewhat higher than would be the case in a perfectly competitive market. Although there is no long-term abnormal profit, there is still some loss of efficiency

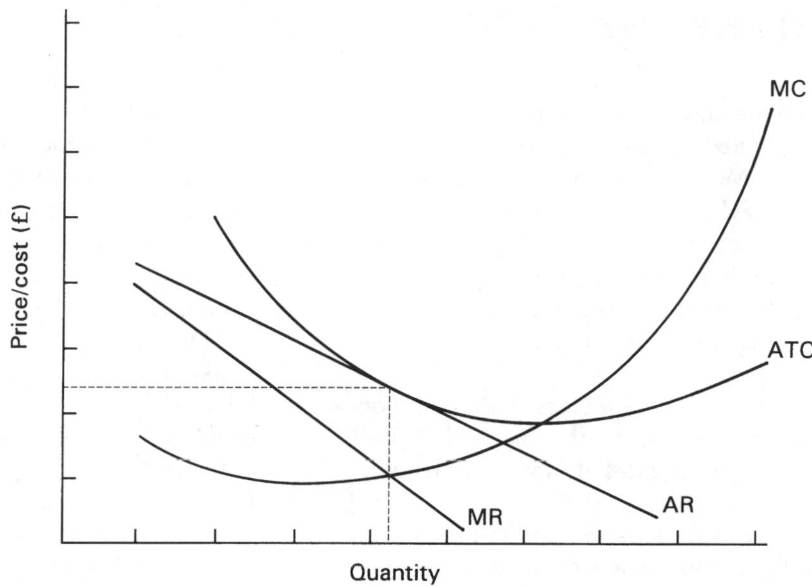

Figure 9.6 Monopolistic competition (equilibrium).

in supplying the market. Unlike perfect competition, monopolistic competition allows for variety and choice, but at some cost to the consumer.

CASE STUDIES

This chapter is essentially theoretical. It examines the conditions for maximizing profit. It is analytic rather than practical and you should concentrate on understanding the different concepts of cost and the relationships between them.

WORKSHOP

1 ***a*** Use the schedules of figures given in the main text to plot accurate graphs of ATC, MC, P, and MR. Assume a price of £22.50 in a competitive market. (For best results marginal figures should be plotted mid-way between production levels, as they represent the cost of moving from one output figure to the next.)

b Compare the outcomes from the consumer viewpoint of competitive and monopoly markets if firms aim to maximize profits.

c If there is a recession and the price falls in the competitive market, at what point will the firm close down production? (You will need an additional cost curve for this – which one?)

2 ***a*** List as many fixed costs as you can associated with a general building contractor.

b List the variable costs for the same business.

c Are there any items which appear in both lists, or which you cannot assign easily to either list?

d What conclusions can you draw about the nature of fixed and variable costs?

3 A contractor with an annual turnover of approximately £10 m has ten contracts, each worth £1 m, over a twelve-month period. The jobs were priced with a mark up of 14% on the total site costs to allow for head office costs, which amount to £1 m p.a., and profits. The jobs have been going well and although one is over budget because of unexpected drainage problems this is compensated by two others which benefited from a spell of good weather. The cost-value reconciliations show the firm is meeting its targets.

The firm is invited to tender for extensions to a local branch of a well-known national retailer. In order to take the job on the contractor would have to hire extra site labour although they would not need to expand their head office operations. Competition is keen and the firm's estimator reckons there is no chance of winning the job unless they cut their prices. He suggests slashing the mark-up to .3% of the total site costs of £1 m.

a What is the usual profit margin allowed for on jobs?

b Using the rules of profit maximization, would you advise the firm to go ahead with this tender?

c Are there any considerations, apart from the impact on profits, which would be relevant to this decision?

d Would your recommendation be any different if the cost-value reconciliations showed the firm was likely to end the year with a small loss on its current workload?

The market for construction work continues to weaken and eighteen months later the industry is in a recession. Forecasts of an upturn in orders have not yet been fulfilled. The firm has laid off a quarter of its directly employed labour force and is struggling to find enough jobs to keep the remainder at work. Currently five £1 m contracts are under way, three are on target (the bids allowed for full costs plus 2% profit), the last two were priced with no allowance for profit. Tender prices are still falling. Bids are being invited for a local authority contract. Surveyors have priced the work at £1 m before overheads and the estimator advises there is no prospect of winning the job at a full cost price. Current staffing levels are fully adequate to undertake the work.

e Assuming current workload is completed to plan, what will the firm's financial position be at the end of the year?

f Advise on whether or not to prepare a tender for the local authority work, and at what price level.

4 A monopoly is defined, theoretically, as the sole supplier to a market. The extent of the monopoly depends on the extent of the market – it may exist at a local, national or international level. Consider a small builder, in a rural area: the only builder in a 25-mile radius, he may be said to have a monopoly of local work.
What is the basis of this monopoly? (What entry barriers prevent competitors from entering this market?)
Are his prices likely to be higher than those of builders working in closer competition with each other?
Is the lack of competition likely to affect the quality of his work? In what way?
What are the limitations, if any, on his ability to exploit his monopoly position?

5*Where there is perfect competition firms compete on price, where there is monopolistic competition firms compete on service and products, and where there is monopoly the client has no choice.
To what extent is the above statement true? Examine each situation from the client's viewpoint, using examples from the construction industry.

6*Outline the characteristics of perfect competition, monopolistic competition, oligopoly and monopoly.
How competitive are construction markets, in the light of your descriptions?

DISCUSSION QUESTION

❑ How useful is MC = MR in developing an understanding of how firms operate?

10 Size of firms

PREVIEW

- Why are some markets (e.g. car manufacture, highway construction) dominated by large firms?
- Why are other markets (e.g. car repair, housing maintenance) full of small firms?
- What are the advantages of large-scale production?
- How do so many small firms survive in the building industry alongside the large firms?

THE SCALE OF OPERATIONS

The analysis of profit maximization in Chapter 9 was concerned with adjustments to output in the short term. The short term is the length of time during which the firm is constrained by its fixed factors of production, things which it cannot readily change such as the amount of factory space, plant and equipment, or its management capability. In the longer term these limits disappear; premises can be enlarged, skills improved and management structures strengthened. The firm's capacity is expanded.

As the scale of operations alters the cost curves looked at in Chapter 9 shift. These were short-run curves based on a particular set of fixed costs. As the firm grows its fixed costs change and the curves must be re-drawn. Each jump in size takes the curves further to the right on the graph. A series of short-run average cost curves (SRAC) can be built up as the firm expands. From these a long-run average cost (LRAC) curve can be constructed (Fig. 10.1).

In most industries the LRAC has the same characteristic U-shape as the SRAC. As the firm moves from $SRAC^1$ to $SRAC^2$ it moves to the right, an increase in the volume of output, and downwards, a reduction in average costs of production. This shows an initial benefit from **economies of scale** – output is rising faster than costs. Then there may be a period of constant returns to scale with output and costs increasing at the same rate, leaving unit costs unchanged. Finally expenditure on costs starts to accelerate, rising faster than output, and **diseconomies of scale** are experienced.

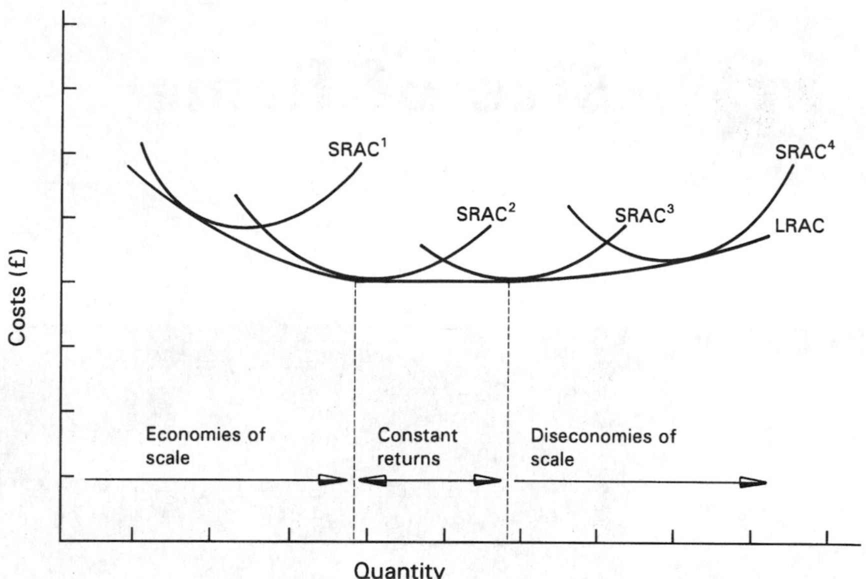

Figure 10.1
Short- and long-run average cost curves.

ECONOMIES OF SCALE

What makes it possible to produce a large volume of goods at less cost per unit than smaller volumes? Many economies of scale are linked to mass production methods, based on division of labour. By dividing work into small tasks, each of which is undertaken by a specialist worker, output can be raised substantially.

A classic description is given by Adam Smith, the founding father of economics, in his account of a pin factory: 'One man draws out the wire, another straightens it, a third cuts it, a fourth points it, a fifth grinds the top to receive the head; to make the head requires two or three distinct operations; to put it on is a peculiar business; to whiten it is another; it is even a trade in itself to put them into paper. The important business of making pins is, in this manner, divided into about 18 distinct operations.'

The outcome of organizing work like this, instead of having each worker making pins from start to finish, is a huge increase in productivity. Adam Smith estimated output to be several thousand pins per man, per day, instead of a few dozen. (The reasons for such a staggering increase are worth thinking about. How much scope is there for organizing construction work in this manner? Consider, too, any drawbacks to this system of working, which finds its epitome in the conveyor-belt production line.)

Production methods are allied to the technology employed. Mechanization was linked to the division of labour; it would have been difficult in Adam Smith's day to devise a single machine capable of turning wire into pins, but once the job is broken down into simple processes, it is relatively easy to mechanize each stage. Automation has in its turn reduced our reliance on repetitive labour. Now we are beginning to return to more flexible ways of working.

Where growth in size changes production techniques a firm may benefit from **technical economies of scale.** An example of different methods is the use of a bulldozer compared to a shovel. The man with a bulldozer shifts more earth than a man with a shovel, but a small firm may not have sufficient work to justify investing in such expensive plant. Mass production techniques generally use more and different equipment than production by hand. Technical economies also arise from scaling up plant or machinery. It does not take twice as many bricks or twice as much labour to build a factory of 10 000 sq. ft as to build one of 5000 sq. ft. A large truck does not require twice as much steel to make it as a smaller truck with half its capacity, nor will it cost twice as much to run. Neither fuel nor labour costs rise in proportion with its cargo capacity. Here we have an example of indivisibility – there is a minimum manpower requirement (one driver) and we cannot scale down the costs for smaller loads. Many costs, especially large plant such as heavy earth-moving equipment or large lifting gear, exhibit a similar 'lumpiness' – they cannot be scaled up or down in line with output.

Technical economies are generally less important in the construction industry than in factory-based industries. Partly because it is site-based, construction does not employ such a large amount of fixed capital. In addition plant and equipment can often be hired which minimizes the need for substantial investment by the firm. This reduces the extent to which small firms are at a disadvantage through their size. When construction moves off-site and starts to use prefabrication techniques the scope for technical economies of scale becomes much greater.

Other benefits of size are linked to **managerial economies.** Management is a major resource for contractors, so larger firms may gain some advantage here. Managers in small firms are involved in every aspect of doing a job and running a business. Larger firms can separate these functions and hire specialist managers to deal with estimating, quantity surveying, marketing, site supervision, personnel, plant, etc. As specialists they acquire a higher level of expertise. Some firms even run their own research and development programme although this is not a high priority for construction firms. Larger firms, who operate in more than one field of activity, usually organize them as separate subsidiaries, one of which may focus on innovation. The Laing group, for example, has an Energy, Technology and Environment Division. Its activities range from investigating soil sampling techniques to load measurement on the historic Menai Straits suspension bridge. Most of its work is done on a consultancy basis but research into matters like energy conservation support other activities in the group such as housebuilding.

Marketing economies relate to buying and selling functions. Bulk purchases of materials and components mean better discounts and lower costs. In addition being a big customer gives the firm greater bargaining power in setting specifications, delivery dates, etc. On the selling side, advertising costs depend on space or time purchased rather than on the volume of goods involved. Distribution is similarly cheaper in bulk.

Financial economies arise in several ways. Reputation makes borrowing easier and having assets to offer as collateral makes it cheaper. Larger firms are often public limited companies, which give them the widest possible access to permanent capital via the issue of shares. Many risks are easier to calculate where large numbers are involved. This eases the burden of contingency funding. A builder who relies on a single contract for a large proportion of his annual workload can find the whole business in jeopardy if that contract goes wrong.

DIVERSIFICATION

Larger firms, as well as running more contracts, can reduce exposure to risks by working in different sectors of the market. In the downturn suffered by housebuilders in the mid-1970s a number of firms such as Tarmac took the opportunity to buy out builders, in this case John Maclean, to broaden their base. Many takeovers are aimed at diversification, either to move from a declining sector into an expanding one or to combine activities with different business cycles.

Acquiring a business by takeover brings instant access to the new market, along with expertise, but a firm can also diversify by internal development. This is slower and in some ways more risky as it will have to build up contacts and experience in an unfamiliar market, but buying an existing company can have its pitfalls too. The main assets of construction companies are the quality of their management and their contracts. Judging quality from the outside is not easy and management skills are easily lost. People often look for new jobs when faced with the uncertainty that usually accompanies a takeover.

One new market emerging from the recession of the early 1990s is facilities management. The term can cover a range of activities, but involves taking responsibility for managing a building, including minor works and maintenance programmes. As firms in all sectors of the economy have sought to withdraw from peripheral activities, opportunities for facilities management have increased. The burden of caring for their own built environment is something which public and private sector enterprises are willing to contract out. One Wimpey manager predicted a move towards total packages offering clients a service from planning and design, through construction to regular maintenance and repairs, and perhaps even extending to matters like security and administration services.

Diversification has been a prominent feature in the growth of larger contracting companies. Subsidiaries in different sectors benefit from the security offered by being part of a larger group. At the same time they can complement each other. Property development or housing, for example, require substantial finance until the building is completed and ready for sale. Contracting, by contrast, generates a cash surplus as the work proceeds. The two provide mutual support.

In some cases companies have expanded beyond the confines of the construction industry. The civil engineering firm, Mowlem, in a series of acquisitions in the 1970s and 1980s, widened geographical spread and product range, moving via construction testing equipment into technology-based products. Tarmac, another company with a broad range of construction-related interests, extended into mining and engineering and even acquired a main dealership for Ford cars.

A wide geographical base helps to reduce risks by limiting dependence on a single region. Housebuilding in the south east of England suffered severe recession in the late 1980s but national companies were less badly hit than regional firms. International operations extend the spread. Taylor Woodrow stated that its 'diverse range of global construction activities enabled it to lift the volume of work in this sector, despite political and economic uncertainties in many territories' (1992 Annual Report). A case of not having all your eggs in one basket.

DISECONOMIES OF SCALE

The bigger the better is not a universal truth. Sheer size can lead to inefficiency. Management becomes more complex and, as the chain of commands lengthens, there is a danger that decision-making becomes less responsive to the market. Administration tends to grow faster than production and sales. This is reflected in rising fixed costs.

The smaller to medium sized firm can operate from a single office. A site foreman supervises the day-to-day work, with regular visits from the quantity surveyor or management representatives. Communication lines are short, and if problems arise they can be referred back to the centre very quickly. With very small firms supervision is in the hands of the manager and decisions are made on the spot.

Larger firms generally operate over a much wider area and need to set up regional offices. Within the regions there may be two or more local offices.

This administrative hierarchy creates additional overheads, but is aimed at maintaining a quick response to urgent day-to-day problems, while matters of more strategic importance are referred back to a higher level of authority. Where a firm accepts a wide variety of work its primary structure may be divided by specialisms, e.g. housebuilding, civil engineering, project management, etc., rather than by region.

The advantages of centralized as against decentralized organization vary with circumstances. Wimpey, for example, was criticized for a top-heavy structure at the end of the 1970s; some reorganization took place but the company had also extended its activities into speculative housebuilding. It was taking on a greater variety of work and needed a relatively high proportion of administrative staff to cope with supervising more sites and coordinating different activities. The overheads were not simply a function of size, they also reflected the nature of the tasks.

Really large organizations often find it difficult to keep in touch with their own grass roots. The labour force may feel a lack of personal involvement where decisions are handed down from a remote head office. If morale suffers, productivity is likely to suffer also. Customer contact is more remote and the firm has to work hard to create a friendly and accessible image. In the construction industry personal contacts play an important role and local knowledge is often valuable in getting tenders and winning contracts. In these circumstances a decentralized structure, based on regions, may be more effective than a streamlined but highly centralized organization based on functions. A company operating overseas has even more problems to tackle. Overseas markets vary from one country to another. Companies will need to develop an understanding of local requirements, and put in place systems to meet local regulations, comply with tax and currency laws, etc.

Many firms which diversified in the expansionary climate of the 1980s found themselves overstretched in the less favourable atmosphere of the 1990s. Selling off the peripheral parts of the business improved their balance sheets and enabled them to concentrate efforts on their core activities. A similar response can be traced in earlier downturns in the market. Higgs and Hill gave up civil engineering work when that sector declined in the 1970s. Costain, badly hit by recession and losses on its Channel Tunnel contract in the late 1980s, announced its intention to focus attention on its core businesses of contracting and mining. Wimpey's annual report in 1991 spoke of 'a demanding programme of disposals of non-core and underperforming assets. The target of £300 m was achieved ahead of schedule with successful sales of the waste management and off-shore fabrication businesses and major property disposals.'

To conclude we can see that growth may bring economies of scale at first, but it also creates problems of coordination and general management, which may lead costs to rise again. Where there is a shortage of particular resources firms may also find the prices of these go up as they increase their demand for them. For these reasons the long-run average total cost curve is likely to resemble the short-run curve in being U-shaped.

It is not inevitable that all businesses will follow this pattern. For instance, in traditional housebuilding there is limited scope for economies of scale. The work is labour intensive rather than capital intensive and the size of the site limits the scale upon which the work is carried out. Growth simply means more sites, it does not greatly change the nature of the job. The resulting cost curve may turn out to be more L-shaped than U-shaped, with initial benefits from increased size tailing off quite quickly and costs remaining fairly stable thereafter.

Evidence is scanty and analysis is complicated by the fact that the growth of large-scale construction firms is entirely a post-war phenomenon. As this has also been a period of far-reaching changes in technology, it is not easy to distinguish the effects of size from the effects of new methods and materials – especially as it is the larger firms which are more likely to adopt the new technology.

SMALL FIRMS

How do small firms survive in the face of the advantages enjoyed by their larger competitors? Industries differ greatly. There are very few small steel producers, car manufacturers or oil refineries but there are many small shopkeepers, builders, hairdressers, accountants, hoteliers and garages. What do these small businesses have in common?

Often they are locally based, serving a small market in geographical terms. They can develop close contacts with their customers. Small housebuilders are often well placed to purchase individual building plots because of their local contacts. Some small firms are specialists, others are tradesmen, such as plasterers or roofers, who work as subcontractors for larger firms. Frequently there is a strong service element, where personal contact is important. A builder dealing with household repairs and improvements is expected to advise and interpret the client's wishes, as well as organize the work. With a small firm the client has the reassurance of dealing with the principal and knowing the work will be undertaken or supervised by them. Modifications to the original plans are more easily handled and problems can be resolved on the spot.

Some small firms are small because they are new. Not all of them stay small. Many will fail, others will grow and may eventually become large empires, but the majority of successful small businesses which stay small do so because their owners enjoy taking personal responsibility for running the firm, and providing a good service for their customers and a friendly atmosphere for their employees.

THE CONSTRUCTION INDUSTRY

Few industries can match the range of firms found in the construction industry, from international giants at one end of the scale to thousands of small firms and self-employed individuals at the other. The majority of firms are small, 95% employ fewer than ten men, but half the industry's workload is undertaken by the top 1% of largest firms. As in other industries, there has been a tendency for the number of firms to decrease and the average size to grow in the 1960s and 1970s, but this trend was less marked in building than in other sectors. In more recent years increased subcontracting and a huge expansion of very small firms has partially reversed the trend, so it is the middle ranking businesses where numbers have declined (Table 10.1).

Small firms are encouraged by the ease of entry into the industry. There are no entry barriers in the form of qualifications or licences, the amount of capital required is small and the widespread practice of subcontracting enables small firms to get involved in contracts much larger than they could handle on their own. Moreover, repair, maintenance and improvement work, especially in the housing sector, means a constant flow of smaller jobs, which are suited to local firms (Fig. 10.2).

Smaller firms have faced considerable pressure in recent years. As the number of large contracts has declined, smaller firms have found their markets under attack from larger companies. In many cases these have set up separate small works divisions to attract jobs which they would not have

Table 10.1 Number of firms and value of work done by size (number of employees)

No. of employees	No. of firms	% of all firms	% change 1981 -1991	Value of work (£m)	% of total	% increase 1981 -1991
1	103 169	49.7	+154	809.3	8.7	417
2–7	92 118	44.3	+68	1638.0	17.7	161
8–114	11 450	5.5	–38	2800.1	30.3	64
115–1199	626	0.3	–25	2621.2	28.3	90
1200+	39	0.02	n.c.	1369.2	14.8	120
All firms	**207 400**	**100**	**+80**	**9237.6**	**100**	**106**

Source: Compiled from *Housing & Construction Statistics*. HMSO, 1993. Reproduced with the permission of the Controller of Her Majesty's Stationery Office.

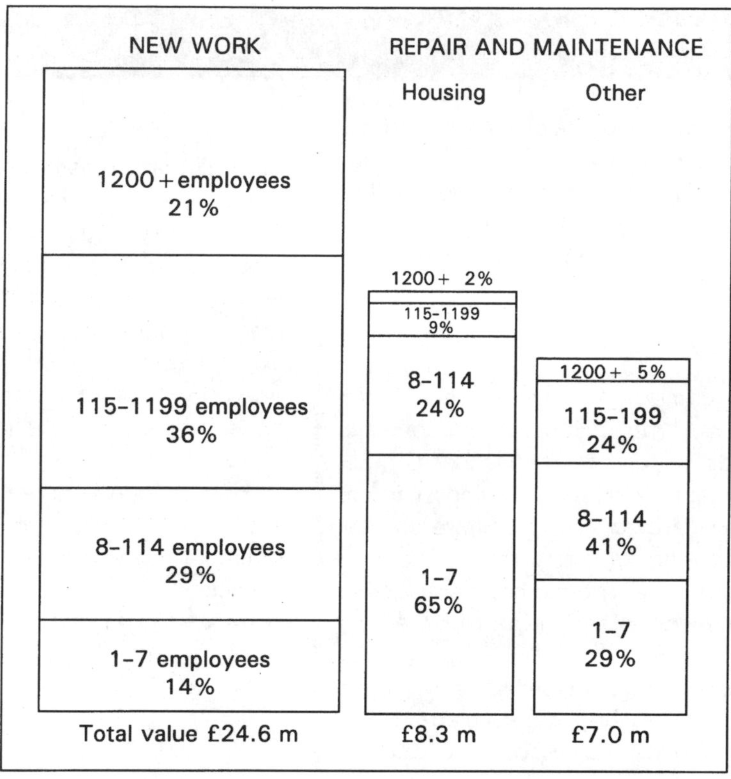

Figure 10.2 Market share by size of firm 1991. (Source: Based on figures in *Housing and Construction Statistics*, HMSO, 1993.) Reproduced with the permission of the Controller of Her Majesty's Stationery Office.

considered worthwhile in the past. At the same time established small firms have found growing competition from self-employed tradesmen who can offer low prices because they have virtually no overheads.

Ease of entry into the industry is matched by ease of exit. Bankruptcy rates are high. The latest recession (early 1990s) has been sufficiently prolonged to see the failure of many well-run small builders, who simply lack the resources to survive a lengthy period without enough work. Diversification at this level is not easy, although some have tried to find work in the public sector, or sought to develop niche markets, such as grant-aided conversions for disabled people.

Scope for cutting costs is also limited. Offices may be simply a part of the family home, and transport double up as personal transport. Labour is subcontracted as needed and stockholding of materials is unlikely to be substantial. A downturn in business soon erodes profit and many small firms that survived did little more than pay their proprietors' living expenses during the recession. In these instances, survival can be put down to the zero opportunity costs of continuing the business, the fact that there were no prospects of finding work elsewhere.

CASE STUDIES

NB: The first case study relates to a contract undertaken by Frankipile in 1972. For this reason the technical details are outdated, but it remains a clear example of how specialization and economies of scale can benefit a company.

A.

As part of a campaign to keep ahead in the piling business, Frankipile continues to refine its well established bored pile techniques. These refinements include mounting a tripod rig on a crawler-tractor; using a more efficient type of clay boring tool; and relying more on its own assessment of ground conditions rather than using the consultant's report as the only basis for pricing work.

One important responsibility for Frankipile is to ensure piles are constructed to the specification. For this the company relies on its own labour under a supervising foreman. The importance Frankipile attaches to keeping experienced labour is emphasized by the fact that when piling work is slack in any area, key operators are paid a retaining wage, which dissuades men from leaving the company as soon as a contract is finished.

The company's nationwide approach has allowed it to build up a library of soils information showing the bearing capacity of various soils throughout the country. Part of their skill in tendering is to relate this information, in conjunction with the consultant's report, to the design of piles. In this case the Kimmeridge clay was strong in compression but weak in cohesion. This meant the end bearing value to the piles was greater than for normal clay, but the shaft was made longer to compensate for the frictional deficiency. The client's confidence was shown by the fact that piles were constructed before the results of the test pile were known.

In all piling operations time is lost as the rig moves from one position to another. To solve this Frankipile has developed its own crawler mounted rig. Work has been going on for two years with the production of several prototypes. A move over 15 m with the old rig took 13 minutes in good conditions. The same operation with the latest crawler is over in three minutes, with the rig set up and ready to drive.

Piling operations of the current contract are expected to finish two weeks ahead of schedule – a worthwhile saving on a 12-week contract!

Adapted from Construction News, *November 1972.*

B.

The embarrassment of missing an industry boom!

Mr Robert (Bobby) McAlpine had the embarrassment recently of trying to explain how his construction company, Alfred McAlpine, had managed to achieve a 35% fall in pretax profits last year when most of its rivals will announce big gains.

Mr McAlpine admits the company misjudged the market. 'We failed to recognize the extent to which investment in private sector building would grow. Instead we continued to chase low margin road and public works contracts.' Losses incurred on a number of major contracts may turn to profits when claims are finally settled.

The group had other problems last year. Big hopes had been pinned on the expansion of concrete block-making operations, but the block-making division underestimated the amount of time it would take to bring new plant on stream. Misjudgements were made about the right product mix between light and heavy blocks. The group has reacted by re-organizing its management and bringing in a new director to run this part of their operations.

The contracting division has also been completely restructured, with the creation of eight new regional offices. Previously the group attempted to run all its work in southern England from Wolverhampton, while contracts in the north were run from head office in Cheshire. 'The regions have become increasingly important ... The lack of regional offices inhibited us from winning as big a share of this work as we would have liked,' said Mr McAlpine.

Benefits from the reorganization are expected to start showing in the second half of the year.

Adapted from an article in
Financial Times, *© F.T., 1 March 1989, reproduced with kind permission.*

C.

Balfour Beatty is an international construction group, based in the UK, with twelve operating companies. Its range of construction activities extends from housebuilding to the construction of power stations and marine cable laying. In 1989, when it embarked on a major expansion of its IT facilities, it had a turnover of £1.6 bn and 20,000 employees.

Its IT network is designed to link the computer resources of its twelve companies so as to exploit the expertise in the group to the maximum. Price was not the main criterion, although 'as a group we can demand fairly good discounts from many suppliers. They know if they serve us well they will get follow-on business.' Choice was based on reputation, adopting the hardware and software known to be tried and tested.

Uses for IT vary hugely. Some sites just need a single PC for basic accounting and word processing. Stock control is much less of a headache than before and coordinating the information from different sites has made for a more efficient bulk buying system for site materials. The databases on component prices and availability can be accessed by estimators from any company in the group. Where the company is hoping for even bigger savings in the future is in networking its very expensive plotters to improve its computer-aided design (CAD) facilities. Already it is standardizing on AutoCAD, which proved invaluable on the contract for Rupert Murdoch's printing plant at Wapping. Originally designed for black and white presses, the plans were changed at the last minute to allow for colour printing. Without CAD the extra work involved would have been horrific.

The group operates an IBM mainframe for accounts and central administration. It has 75 minicomputers and over 800 workstations. PCs operate on local networks and a third can now access central IT resources. Soon they will have one terminal for every three to four staff. It is a strategy being 'driven by commercial necessity' according to the group's IT manager. 'Recession focuses the mind enormously' and IT is helping to keep projects on time and within budgets.

Reproduced with permission from:
Which Computer, *February 1991.*

WORKSHOP

1 ***a*** Give examples of different types of economies of scale experienced by Frankipile (case study **A**).
b What sort of economies of scale do you think are available to Alfred McAlpine (case study **B**)?
c How does IT benefit larger companies (case study **C**)?
d How important were economies of scale in the merger that created Hall and Tawse (case study **C**, Chapter 8)?

2 To what extent would you ascribe McAlpine's difficulties to diseconomies of scale?
Can you suggest circumstances in which Frankipile might experience falling profits? Would the scale of its operations be helpful or add to the difficulties, in these circumstances?

3 ***a*** Would you categorize your own firm as small, medium or large?
b What do you consider to be its main strengths?
c How far are these related to its size?
d What do you think is its main weakness?
e Can you suggest a strategy for improving this?

4*Describe the main economies of scale experienced by a large construction company.
In view of these advantages, how do you account for the existence of so many small firms in the industry?

DISCUSSION QUESTIONS

In the past two decades the construction industry has experienced two major recessions as well as prolonged periods of prosperity.

❑ Do you consider small firms or large firms more likely to survive a slump in the industry?

❑ In a boom period, which will be better placed to reap the benefits of high levels of demand?

11 The construction industry

PREVIEW

- What is 'the construction industry'?
- How large is the industry and how can it be measured?
- In what ways does the construction industry affect the rest of the economy?
- How does construction compare with other industries?

Before going on with economic theory it is time to take a closer look at the construction industry itself. It is a large and diverse industry. The Standard Industrial Classification, which defines industrial categories for the purposes of all official statistics, lists over twenty separate trades under the heading of construction. The list includes:

> ... erecting, repairing buildings; constructing and repairing roads and bridges; erecting steel and reinforced concrete structures; other civil engineering work such as laying sewers and gas mains, erecting overhead line supports and aerial masts, open cast mining, etc. The building and civil engineering establishments of defence and other government departments and of local authorities are included. Establishments specializing in demolition work or in sections of construction work such as asphalting, electrical wiring, flooring, glazing, installing heating and ventilation apparatus, painting, plumbing, plastering, roofing. The hiring of contractors' plant and scaffolding is included.

Workers employed by firms outside the industry to undertake building jobs, e.g. the works department of a manufacturing company, are classified by their employer's main business. On-site industrialized building counts as construction, but factory-based production of prefabricated units is classed as manufacturing. The majority of construction firms engage in relatively little non-construction activity, so these minor anomalies mean that official figures will tend, if anything, slightly to underestimate the role of the industry.

A STATISTICAL SNAPSHOT OF THE INDUSTRY

Although construction has declined in size from its peak in the late 1960s, it is still very substantial, contributing approximately 7% of the gross domestic product. As we entered the 1990s, new building accounted for 58% of work done, the other 42% being repairs, maintenance and improvements. A third of all work was in the housing sector. A total of 156 995 new houses were completed, a considerable drop from the peak year of 1968 with 414 000 completions and 23% less than 1980. Overall less than a quarter of new work was for public authorities, with more than three-quarters for the private sector; this represents a considerable shift away from the public sector (Table 11.1).

In the mid-1980s a third of all new work was for public sector clients, but it must be realized that some of these clients, such as the gas, electricity and water industries, have themselves become part of the private sector. In housing the public sector share of newly completed houses fell to under 10% of the total, compared with 36% in 1980 and 46% of completions in 1968. The shift in favour of the private sector and the predominance of owner occupation has been a feature of the past decade (Table 11.2).

Construction work contributed just over 50% of gross domestic fixed capital formation in 1990 (this includes housing). The industry provided approximately 1 m jobs, nearly 5% of the nation's employment, with more than 20% of self-employed persons engaged in construction work.

These statistics show the industry plays a significant role in the economy. It provides a large part of the nation's infrastructure, its roads, power stations, public buildings, etc. It is intimately involved with economic growth and prosperity. It is a major supplier of investment goods to commerce and

Table 11.1 Orders for new work obtained from public sector (£m)

	Housing	Transport related work	Education and health	Other	All
1981	672	1133	759	1842	**4361**
1982	984	947	718	1767	**4416**
1983	985	1057	804	2313	**5159**
1984	876	1171	988	1991	**5026**
1985	734	1106	833	1939	**4612**
1986	772	1333	794	2015	**4914**
1987	903	1331	861	2321	**5416**
1988	882	1301	1213	2602	**5998**
1989	872	1668	1435	3092	**7067**
1990	683	1675	1336	2136	**5830**

Source: Compiled from *Housing & Construction Statistics*, HMSO, 1993. Reproduced with the permission of the Controller of Her Majesty's Stationery Office.

Table 11.2 Orders for new work obtained from private sector

	Housing	Industrial	Shops and offices	Other	All
1981	2013	1554	1982	812	**6360**
1982	2928	1327	1935	837	**7028**
1983	4078	1543	1758	980	**8358**
1984	4001	2203	2303	877	**9604**
1985	4555	2149	2797	1230	**10732**
1986	5421	1993	3358	1423	**12195**
1987	6441	3660	4726	1876	**16703**
1988	7894	3128	6633	2647	**20301**
1989	6497	3377	7357	2836	**20066**
1990	4855	3736	5561	2510	**16663**
1991	4552	3452	3438	2374	**13815**

Source: Compiled from *Housing & Construction Statistics*, HMSO, 1993. Reproduced with the permission of the Controller of Her Majesty's Stationery Office.

industry affecting the quality and convenience of the workplace. It creates jobs and incomes for a great many people and operates in all parts of the country.

A large part of the industry is concerned with housebuilding and maintenance, which has a direct impact on our standards of living, as well as providing a springboard for consumer spending on furniture, household appliances, DIY goods and garden products. It is not an exaggeration to say the built environment in which we live and work affects nearly every aspect of our lives, our productivity, our comfort, and our cultural and aesthetic sensibilities.

A sizeable, but dwindling, part of the industry's work is for the public sector, which makes it sensitive to changes in government policies. In recent years cuts in government expenditure have meant fewer contracts and falling employment for construction workers in the public sector. High interest rates in the late 1980s extended the cutbacks to the private sector. The building boom turned into a recession, with the industry experiencing falling profits, falling employment, falling output and rising bankruptcies.

CONSTRUCTION AND OTHER INDUSTRIES

Construction firms are involved in relatively little non-construction activity. At the same time only around 5% of construction output comes from firms not part of the industry, mainly building maintenance by directly employed staff in other industries. The category of 'construction' is therefore a well defined class with little overlap into other activities. Nonetheless, it has many links with other industries, whether as customers for construction work or suppliers to it (e.g. steel). The extent of this interdependence can be measured by input–output analysis.

Input–output data are compiled from census returns which firms are obliged to complete. The basic tables show the value of sales of each product, classified by the industry which produces it. The commodity 'construction' is thus the principal product of the construction industry.

The tables can be used to analyse the flow of inputs required by one industry, e.g. construction, from other sectors of the economy. Similarly, sales to other industries and to final consumers can be traced. By looking at transactions between industries we can examine the links between them and the impact of changes in one sector on others.

The analysis of purchases by industry, an extract of which is shown in Table 11.3, shows the various inputs used in the columns. Construction, no. 88, makes no direct purchases from agriculture, but it buys £6 m of forestry products and £2 m of coal. The largest input is from itself! This shows construction firms are constantly buying in materials and services from other firms in the industry. All the inputs, plus imports, wages, salaries and taxes, give that industry's outgoings, the total at the bottom of the column. Construction sales to each industry and to final consumers are shown along the row no. 88.

Another type of table, based on these statistics, looks at the total inputs required by each industry to produce 1000 units of final output (Table 11.4). This approach emphasizes the extent to which industries are interdependent. For example, in 1984 direct purchases of coal by the construction industry amounted to £2 m, or less than 0.005% of its total

Table 11.3 Extract from 1984 input–output statistics: Purchases by industry from domestic production (HMSO, 1988). Reproduced with the permission of the Controller of Her Majesty's Stationery Office.

Sales (£m) by commodity group	**Purchases by industry (£m)**			
	86 Plastics	**87 Other manufacturing**	**88 Construction**	**89 Distribution etc.**
1 Agriculture	—	1.0	—	165.0
2 Forestry & Fishing	—	0	6.0	3.0
3 Coal extraction	2.0	—	2.0	—
. . .				
86 Plastics	339.0	109.0	211.0	358.0
87 Other manuf'g	0	26.0	61.0	200.0
88 Construction	8.0	5.0	8 841.0	703.0
89 Distribution etc.	101.0	70.0	874.0	1 410.0
. . .				
102 Total input	**4557.0**	**2092.0**	**42 745.0**	**52 822.0**

Table 11.4 Extract from 1984 input–output statistics: Total requirements per 1000 units of domestic commodity output in terms of gross output (HMSO, 1988). Reproduced with the permission of the Controller of Her Majesty's Stationery Office.

Total requirements per 1000 units of domestic commodity output in terms of gross output

	86 Plastics	**87 Other manufacturing**	**88 Construction**	**89 Distribution etc.**
1 Agriculture	1.4	3.4	1.2	8.3
2 Forestry & Fishing	0.7	0.7	1.9	0.7
3 Coal extraction	6.1	4.3	16.9	3.7
. . .				
86 Plastics	1086.3	60.3	10.7	12.0
87 Other manuf'g	1.5	1018.6	3.8	6.1
88 Construction	6.5	8.0	1265.0	24.2
89 Distribution etc.	41.0	51.9	47.9	1044.3
. . .				
102 Total	**1797.8**	**1699.6**	**2077.5**	**1755.8**

purchases of £42 745 m. Construction does not appear to rely much on the coal industry. But overall the demands made by construction on the coal industry amounted to 6.9 units per 1000 units of final output. This is because buying steel or bricks involves indirectly buying the coal needed to make them. The interdependence of coal and construction is thus greater than the basic input figures suggest.

Although construction output is needed by almost all other sectors, many of the industry's sales represent investment by other firms. It is spending for future rather than for current needs. Often this is seen as deferrable because, unlike spending on materials and labour, it is not contributing to current output. Buildings are very durable and the internal space can be adapted and used more intensively to meet short-term needs. Users are apt to 'make do', and postpone new construction if more urgent needs arise. Even repair and maintenance work may be put off when money is tight.

In the long run construction is vital to other industries. It is the major supplier of the country's infrastructure, and without efficient roads, docks, etc., the whole economy suffers. But in the short run the dependence of other activities on the construction industry is low. This places the industry in a weak bargaining position *vis-à-vis* many clients, including public authorities, and makes it vulnerable to reduced workloads during recessions, and damaging cutbacks when the government is trying to restrict public sector spending.

In many ways construction is unlike other industries. The product is usually large, it has a long lifespan, it is not moveable, it is often produced to individual specifications and costs a lot of money. These things affect the way the industry is organized and the technology it employs. The construction world does not look quite like our model of perfect competition, but the market forces are the same. The rest of this chapter outlines the differences; the next chapter examines how these affect the operation of the price mechanism in construction markets.

DEMAND FOR CONSTRUCTION WORK

Compared to many products, the demand for new work is low relative to repair and maintenance. Because buildings have a long lifespan, two or three generations or longer, there is a large existing stock which dominates the market (see Chapter 4). The many older properties also means there is much renovation and repair work, mainly individual jobs. This limits the application of mass production methods.

In many cases new buildings are also designed to individual specifications. Clients have different needs and ideas which influence their requirements. Each site is unique and presents its own problems. Aesthetic considerations also militate against widespread standardization and mass production. Nonetheless, the high cost of traditional craftsmanship, combined with new technology, have pushed the industry into developing industrialized methods of building. Greater standardization in design and especially in the manufacture of components can achieve substantial savings. Despite this British designers still make use of many individually designed elements. Mass production need not mean total uniformity; many housebuilders use a limited number of standard designs, with minor variations in cladding materials, decorative brickwork patterns, etc., to give each house an individual appearance.

High prices are another characteristic of buildings. Construction work, especially housing, generally represents a large outlay in relation to the customer's income. Most purchases are made with borrowed funds and so the cost and availability of credit are major influences on demand. Fluctuating interest rates can lead to instability in property markets with prices and profits affected accordingly. Building appears to suffer from more severe cyclical movements than many other industries. (Recent empirical studies have cast some doubts on this view, but the recession of the early 1990s has certainly been exceptionally severe in the building sector.) Economic growth generally is far from stable and trade fluctuations seem to occur on a 8–10 year cycle. As a capital goods industry, construction is subject to accelerator effects which magnify these ups and downs (see Chapter 21). In addition a separate cycle of approximately 15–20 years' duration has been noted in housebuilding, both in Great Britain and in other countries.

Government policy has considerable influence on the industry both directly and indirectly. Changes in taxation and subsidies, such as the imposition of VAT on home improvements, can affect demand. Subsidies, in the form of tax relief, did much to encourage the trend towards home ownership which was such a prominent feature of the 1980s. More directly, the substantial sums spent on public sector contracts, such as schools, roads and urban renewal projects, mean that the industry is very vulnerable to changes in policy on public spending.

Changes in interest rates or public spending programmes are part of policies which are intended to achieve broad economic objectives, not specifically aimed at the construction sector. The government may attempt to control inflation, or to reduce unemployment, by varying capital expenditure programmes. Because these frequently involve construction work the industry has been described as an economic regulator. Construction contracts act as a tap to regulate the flow of expenditure by government. Whatever the benefits to the economy generally, this creates a damaging instability within the industry. However, because major contracts take several years to complete, the flow of work in the pipeline helps to smooth out variations in demand as a result of policy changes.

The industry is widely dispersed throughout the country. As construction work largely takes place *in situ*, production cannot be centralized. With jobs ranging from simple repairs to the building of specialized structures like nuclear power stations, there is also considerable fragmentation of the market by product. Competition is imperfect and construction is best thought of as a large number of overlapping markets, determined by the type and location of the work, rather than as a single market. The size and variety of markets and the difficulty of keeping an up-to-date knowledge of what is happening in them affords some protection from competition. This is balanced by the freedom of entry which ensures there are always new firms eager to find work.

SUPPLY SIDE CHARACTERISTICS

Construction, as noted above, takes place where it is to be used. This requires mobility of labour, with workers being prepared to travel from site to site and repeatedly to look for new jobs as the current site reaches completion. There is a similar need for mobility of capital (plant). It is also necessary to provide facilities to assemble and store materials and afford protection from the weather. Site security is an increasing problem.

There is a low degree of mechanization in the industry, even though much unskilled labour, such as earth-moving, has been mechanized. It has been calculated that whereas manufacturing industry operates with a ratio of capital to output of more than 3 : 1, in construction it is less than 1 : 1. A major factor is the nature of the work. Because tasks follow in sequences rather than being repetitive, mechanization is not an obvious solution. Standardization is still limited and sites vary in ease of working. Consequently the industry remains labour intensive. Labour costs can be up to 50% of the total cost of a job, although the industry is changing. New technology and new materials have been introduced, and more use is made of prefabricated modules. Information technology is also having an impact on design and documentation processes.

Construction is an assembly process. A large number of different products and tasks are involved, which require a varied and adaptable workforce. Many specialist skills are used but there is also a substantial demand for semi-skilled and unskilled labour. On the professional side, too, there is a range of distinctive qualifications, from architect to engineer, distinctions which can hinder communications and impede teamwork. The traditional divide between design and construction inhibits mutual progress. At its most extreme this results in designs which are impossible or very expensive to realize. Initial designs for the Sydney Opera House were considered unbuildable by contractors but the splendour of the final outcome should remind us that imagination need not always be subordinated to practicality and economics!

High unemployment has always been a problem in the industry. Discontinuity of work, the need to move from site to site to find jobs, bad weather and the periodic recessions to which the industry is subject, all contribute to a high proportion of casual labour and high levels of unemployment. To some extent a reserve pool of labour is useful as it reduces the risk of work being held up because of particular skills shortages but too many unemployed workers represent wasted resources.

Vertical disintegration, i.e. the splitting up of work between separate suppliers, is very typical and increasing the growth of specialist subcontractors. These are often quite small firms which lack weight in their dealings with main contractors. As a result they find it difficult to negotiate improvements in working conditions or terms such as payment times. One estimate suggests 90% of site work is now carried out by specialists. The growing numbers of self-employed and very small firms in the industry are a part of this trend. Other reasons for the large number of small firms were examined in the previous chapter. Fragmentation of this sort creates a diffusion of management responsibilities which affects the organization of production and adds to the main contractor's role as coordinator.

CASE STUDIES

A.

Any attempt at forecasting the prospects for building and civil engineering has to begin by dividing the industry into two.

Building, which is largely dependent on private spending, and civil engineering, which is traditionally dependent on public sector spending, behave in different ways, and are likely to continue doing so. Over the past few years, building – particularly building offices, shops, factories and houses for sale – has boomed, while the traditional businesses of civil engineering, such as roads, dams, bridges and sewers, has slumped.

The National Council for Building Materials Producers (BMP) predicts there will be a 4.5 per cent increase in private housing output in 1987 to follow this year's growth to 175,000 new private sector starts, the highest level since 1973. BMP is also predicting a 12% growth in new commercial building. This follows the 12% increase in 1986 in the industry's output of new offices and shopping centres.

Compared to the generally optimistic outlook for the private sector, the outlook for sectors dependent on public spending remains gloomy. The local authority house building programme has fallen to its lowest level for years and the volume of work in other building sectors dependent on public funds is likely to remain static.

Adapted from an article in Financial Times, © F.T., 5 January 1987, *reproduced with kind permission.*

B.

The recession in construction is likely to last well into next year, according to the National Economic Development Office. It forecasts that total UK construction output will fall by 3.5% this year and the same next year.

The largest falls are expected in the private residential and commercial property markets, which have suffered because of the rise in interest rates. Housing starts by private builders are likely to fall to 135,000 this year, compared to 168,000 last year and 216,000 the year before.

Commercial output would fall by only 2% this year, but larger falls could be expected next year. Numerous developments started before the market fell had still to be completed.

Civil engineering was one of the few areas expected to show consistent growth. The water and power industries, both part of the privatization programme, were poised to undertake investment programmes. Public sector non-housing output was also expected to grow by approximately 3% next year.

Adapted from an article in Financial Times, © F.T., *16 June 1990, reproduced with kind permission.*

WORKSHOP

1 How far does it make sense to discuss 'the construction market' as a single entity?

2 Compare and contrast the two sections of the industry (building and civil engineering). Which is the more sensitive to fluctuations in the economy?

3 Suggest reasons for the cyclical nature of the construction market, boom followed by slump.

4 How does the industry accommodate fluctuations in demand?
Can you suggest any strategies for construction firms to minimize the impact of such variations?

DISCUSSION QUESTION

- ❑ Attention focused on the construction industry as a principal sufferer in the recession of the early 1990s. Many commentators also saw the industry as the key to economic recovery. To what extent does the rest of the economy depend on the construction industry?

12 Competition and pricing in the construction industry

PREVIEW

- How does someone wanting a car choose which model to buy?
- How does someone wanting building work choose which builder to approach?
- Is price the main form of competition in building markets?
- Is competitive tendering a guarantee of getting 'value for money'?
- How else can clients procure the construction services to best meet their needs?

In the model of a competitive market with which we started, it was assumed that a uniform product was offered throughout the market and there were many buyers and sellers, who were all well-informed about what was available in the market. In consequence a single price prevailed, the equilibrium price.

The reality of the construction world is rather different. There are certainly a great many firms, but because the market is so fragmented they are not all in direct competition with each other. Smaller firms operate locally, larger firms extend over a wider area and deal with a variety of work. The product, far from being uniform, embraces everything from replacing a fallen roof slate to constructing the Channel Tunnel. Some contracts are so complex that there are relatively few firms qualified to undertake them. 'Joe Bloggs' of Birmingham is not competing for the clients served by 'Sam Smith' of Southend and neither of them is in competition for the same contracts as 'Construction Services International'.

Customers have imperfect knowledge of the market. They may be inexperienced, first-time buyers of construction services. Firms vary in their skills and expertise but buyers cannot walk down the high street to compare products and prices in shop windows. In short, knowledge is hard to come by and competition less than perfect. Where each job is unique, the only way to achieve direct competition is to ask a number of firms to give estimates. **Competitive tendering** is widely used, especially for larger public sector contracts. Local authorities are required to put a large proportion of their work out to tender. In the past open tenders have been the norm – in recent years different systems of selective tendering have become more common.

PRICING BY TENDER

Where open tenders are used the job is advertised and enquiries invited from any interested firms. Specifications of the contract are sent to all who apply. As the specifications have already been drawn up, tenders are considered purely on price; design features are not normally involved. In practice the lowest price is not always accepted, although there must be a valid reason for refusing it. In exceptional circumstances some variations on specifications may be offered; in practice this rarely happens, although the specifications on the contract are not always adhered to.

This system is the nearest approximation to perfect competition and has the advantage that the client can compare prices for what should be identical work and feel confident of getting value for money. Unfortunately there is no guarantee that all the contractors are equally competent, or that their enthusiasm to get the job may not lead to under-pricing and subsequent difficulties in completing the work. When firms are competing hard to get work, low tender prices encourage cost-cutting, which means cheaper materials and components may be substituted for those specified in the contract. Although this does not necessarily affect the performance of the building the client is not getting the agreed quality.

The tendering process itself is expensive and in the nature of things the majority of tenders, 80% or more in open competition, will be unsuccessful. In recessionary times, when clients can 'shop around', the success rate will be very much lower. In 1992 a survey among small builders in south Hampshire found, despite a shortage of work, many of them were suffering from too many enquiries. They were kept busy preparing estimates but the number of jobs that resulted was very low. The costs of estimating must either be absorbed by the builder – in small firms it often results in the proprietor working unpaid overtime – or it has to be allowed for in the overheads, making open tenders an expensive system to operate.

Selective tenders partially avoid these problems while holding on to the competitive element. Only firms of known ability are invited to tender. Because the firms have already been vetted, there is greater control over quality and the invitation list can be kept quite short. This reduces the cost of sending out specifications and limits the number of unsuccessful bids. Unfortunately some clients have taken advantage of contractors' eagerness for new work to draw up 'select' tender lists of quite excessive length. One local council advertised a £4 m contract, obtained over 200 replies and invited 156 firms to tender. Bids were received from 111 firms, an absurd waste of effort. The Audit Commission, in its report *Realising the Benefits of Competitive Tendering* (1993), condemns such practices and recommends no more than eight companies be invited to tender in normal circumstances.

In the case quoted above 'selection' was minimal; in reality it appears closer to an open tender. The more usual procedure is to keep a list of approved firms and invite a limited number of applications for specific jobs. Lists must be regularly revised to remove any firms who are failing to maintain the required standards. At the same time newcomers who have acquired the appropriate skills and experience can be added to the list. Firms are competing on quality and reputation to reach a position where they are invited to tender. They then compete directly on price. Selection procedures vary. It may be by rota, by reputation for particular types of work, or by some other criteria. Because competition is no longer open, there is always a danger of firms getting together to keep up prices, especially if a rota system is used whereby firms are always in competition with those next to them on the list (see Chapter 7).

ALTERNATIVE METHODS OF PROCUREMENT

Tendering is meant to ensure competitive pricing, but it does so at a cost of keeping separate the design and the construction processes. Designs must be completed before builders are able to tender, so they can price the work properly. However, it is often possible to improve efficiency by having more consultation between designer and contractor, especially where new techniques are being developed or specialized skills required.

If a particular firm is known to be suitable, the client may opt for a **negotiated contract** with that firm. An element of competition is introduced if two or three firms with suitable experience are invited to discuss the project and an approximate bill of quantities provided to give a guide to prices. Final selection in this process of **two-stage tender** can be made on a combination of criteria rather than simply the lowest cost. On large projects a form of **serial contracting** may be adopted, whereby, provided the firm completes the initial stage satisfactorily, it can expect successive contracts for later phases of the project. This keeps together a successful team and provides continuity of employment. Familiarity with the work helps to reduce snags and improve efficiency.

Separation of design and construction is completely done away with in a **package deal**, where the contractor offers an all-in service for a single price. It is most commonly used in situations where a limited choice of design is acceptable, e.g. light industrial developments. A package deal offers time savings and the advantage of a known price from an early stage. These are important benefits to industrial or commercial clients planning their own production schedules. From the builder's point of view, **design-and-construct** contracts allow the firm to work to its own strengths. The client's brief outlines the basic requirements and firms are invited to submit designs and tender on their own specifications. Competition does not rest solely on price but includes design, methods of construction, etc.

This puts the builder into a position more like the manufacturer who takes responsibility for design as well as production. Some contractors have even developed patented technologies over which they have monopoly rights, such as Wimpey's No Fines or Laing's Easiform systems of concrete construction, both widely used in public sector housing in the post-war period. In general the scope for designing totally new systems which are both practical and structurally sound is limited.

During the 1980s a growing number of major construction schemes, of a size or complexity which would stretch the resources of a single contractor, were successfully organized through **management contracting**. In its purest form this places contractors in an entirely different role from their traditional role as entrepreneurs. The project is no longer an opportunity for profit, for better margins negotiated at the client's expense. Instead, the contractor agrees to manage the contract for a fixed fee (the system was initially known as **fee management**) and becomes a professional adviser to the client. They now sit on the same side of the negotiating table.

There are many variations in the way work is organized and an equally varied array of terminology. Subcontractors may be employed by the management contractor or they may hold their contracts with the client, but remain unanswerable to the **professional construction manager**. Whatever the title, the contractor acts for the client in negotiations with subcontractors, but does not own the resources used to build for the client. The essential features in the many variants which exist are that the contractor undertakes: 'the overall planning, control and coordination of a project from conception to completion aimed at meeting a client's requirements and ensuring completion on time within cost and to required quality standards' (CIOB, *Project Management in Building*, 1982).

Management contracting can lead to considerable time savings, not least because site work can begin before designs are fully worked out. Although the contractors are not responsible for the design process they are involved in consultations with the design team from early on in the project. Where land values are very high an early start means savings in interest charges alone can be substantial. For commercial clients there is the added benefit that the sooner the building is commissioned, the sooner they start to make an income from it. The growth in commercial work during the 1970s, together with the increasingly complex nature of larger projects, encouraged the trend in favour of management contracting.

The system was widely adopted for developments in the City of London during the 1980s and traditional contracts virtually disappeared from this market. Along with changes in contractual relationships came the technique known as **fast-track** construction. Originating in the American 'can-do' approach combined with a design philosophy that aimed to telescope the design and construction stages it produced valuable time savings.

A recent survey showed general contractors expected the increase in management contracting to continue. The majority of larger contractors now offer management contracting services and Bovis, the pioneer in this country, does the bulk of its work in this way. Not all firms welcomed the trend. New methods demand new skills. A good knowledge of the construction industry and leadership qualities, the ability to manage, are more important than specific technical expertise which can be provided by subcontractors. Some managers prefer to use the term trade contractors, which indicates the shifts in relationships implied by this system.

Yet another variant is the appointment of a **project manager**, an approach which has become popular in recent years again because of the increasing size and complexity of jobs. A project manager's function is to coordinate the work for the client, to provide a focal point for communications. For inexperienced clients the help in formulating their requirements is invaluable, from the earliest stages of setting out the brief, through the decisions about what type of procurement to use, to acting on their behalf in consultations during the construction period.

CHOICE OF SYSTEM

Which system is preferred depends on the circumstances of the contract. Traditional methods are well understood; the builder has a contractual relation with the client, but works under the supervision of the client's agent, a role usually undertaken by the architect. The breakdown of these established boundaries leaves some uncertainty as to who bears what responsibilities and the risks that go with them. Where subcontractors' agreements are made directly with the client the management contractor appears to shed all liability for poor performance. The dangers inherent in this, for instance the risk of cost over-runs, can be countered by setting a target price. This gives the management an incentive to keep control of costs because any savings on the target means a bonus, but exceeding the target incurs a penalty.

Design management is another key issue as design work and the 'buildability', or ease of construction, are major factors in meeting the cost and time schedules required. Where supervision is in the hands of a contractor rather than an architect, the practicalities of the building process come to the fore. On the whole clients who use project managers and/or non-traditional procurement methods seem to have fewer complaints than those who rely upon architects to supervise traditional contracts. Complaints centre on high costs and communications failures.

Both problems are tackled by the technique of cost-value analysis, something pioneered by management contractors. The aim is to put together a small team of professionals of different disciplines, perhaps a designer, quantity surveyor and builder, who are briefed to evaluate specifications. They look for alternatives which give cost or quality advantages, reduce waste, or allow for easier construction or maintenance. Variations introduced at later stages are similarly evaluated. With accurate costings and quick decisions, this multi-skilled approach can provide management with a valuable tool in running the project efficiently and clients with value for money.

Larger developments benefit most from a unified approach to management. Such projects exhibit some or all of the following features:

- they are major contracts in terms of value;
- the work involved is complex (the cost per sq. m is high);
- they require an intensive schedule because time is at a premium;
- they use new techniques and/or materials;
- the resources needed are in short supply;
- a large number of people, whose interests and activities must be coordinated, are involved.

The best way of commissioning work depends on the client's priorities. Time and cost are usually the main priorities, but which one is the more important? Where there is a fixed budget certainty of costs may be more important than small savings. Is there a likelihood that they will want to introduce changes after the work has started? How important is appearance – is it a prestige job or in a sensitive location? Finally, how experienced is the client? Traditional procurement gives professional control into the hands of an architect. Newer methods give the client greater involvement but correspondingly greater responsibility for the successful operation of the project.

COMPETITION IN CONSTRUCTION MARKETS

How competitive is the building market overall? Price determination varies from the direct competition of open tenders to various types of negotiated arrangements through to **turnkey contracts**, where the buyer deals with a single firm who will organize the project from start to finish, including site acquisition, design, construction, even finance and fitting out. Only the open tender is close to the model of perfect competition in which suppliers offer an identical product (determined by the contract specifications) and compete on price alone. In other cases monopolistic competition is a better model. This allows for product differentiation with suppliers competing for custom on various fronts – quality, service, speed, etc.

Negotiated contracts might seem to be best described as bilateral monopolies (a single buyer and single seller) but the client can always cease negotiations and invite another contractor to undertake the job. It is a **contestable market**, meaning that there is a potential for competition even though the firm may not be facing actual competitors at present. There are no significant barriers to entry, so should prices be pushed too high the client will look for alternatives.

The client's choice of contractor has usually been made within a competitive framework, although there may be occasions where client ignorance, or the specialist nature of the work, means only one firm is considered. Whether the price that is finally negotiated turns out to be high or low will depend on the bargaining skills of the two sides and economic circumstances of the day. If the client is inexperienced, or perhaps badly advised, or if the contractor sees that the client is already strongly committed, or under pressure to get the work completed, the balance of power is in the contractor's favour. But if the contractor has a very thin order book, or hopes to win follow-on contracts, or sees the job as establishing the firm's credentials in a new market, they will try to keep prices low. Even though monopolies are generally regarded as one-sided, clients who work regularly with the same contractor can build a relationship which is mutually rewarding. Establishing a good understanding of what is required can help contractors to keep costs down and give clients confidence that the job will meet their specifications.

In general the market is competitive. Uncompetitive pricing is mainly due either to ignorance or to market conditions. However good the informal channels of communication, particulars of prices are not public knowledge and neither contractors nor clients can have anything like perfect knowledge of the market. Conditions are variable, the industry experiences considerable ups and downs in demand. In a boom period there may be more work available than firms can manage. In this case not even tendering can ensure keen pricing as firms may prefer to put in excessively high bids rather than refuse an invitation to tender. Alternatively work is 'shared' out by agreement between firms, who act cooperatively rather than competitively.

Not all construction work is undertaken for a client. Speculative builders work entirely at their own risk. The builder finds the site, obtains planning permission, arranges the finance and erects the building. It is entrepreneurial activity, requiring good judgement of the market as well as good management of the work. This is more like manufacturing in that production takes place first and a customer is sought afterwards. Mostly it is confined to the housing sector. Many smaller firms find speculative housebuilding a useful supplement to contract jobs, as it helps them to create continuity of work and plan their production more efficiently.

Risks can be considerable because house prices are quite volatile and development, even on a small scale, takes some time. In periods of buoyant demand the rewards can be equally great. Contractors who moved into development in the boom period discovered how quickly profits can turn to losses when the market collapsed in 1989–90. Keen pricing is closely linked to the state of the market.

CASE STUDIES

A.

Shattered rules, startling innovations and altered attitudes surround the £500 m Broadgate development. The man at the centre of all this is Ian Macpherson, appointed project director to head the management team for Bovis Construction.

Across the whole site, traditional construction approaches are being replaced with what Macpherson describes as 'American-style' building techniques. 'In the US they build faster than anywhere else in the world; or at least they did until Broadgate got under way.' What is the secret? 'You soon realize it is all a question of attitude.'

The Americans work together to get the job done, everyone from client down to smallest subcontractor shares the same goal. 'Compare this,' says Macpherson, 'to the traditional contract arrangements in the UK, where each subcontractor tries to remain autonomous. There is an air of secrecy and a kind of "us and them" attitude ... when someone has a problem, it is *their* problem ... Such an attitude is bad, very bad for the project overall.'

With the emphasis firmly on teamwork, selection procedure for trade contractors is rigorous. Trade contractors – not subcontractors, even the terminology emphasizes teamwork – are interviewed before being invited to tender. Those chosen are asked to look at the work schedule and indicate how they intend to approach it and who will be made responsible. Once appointed trade contractors can look forward to contracts on subsequent phases, again a reflection of the importance placed on a committed team. Of course, performance must be satisfactory too!

There are three basic elements in controlling a project: design, cost and quality. All members of the team must cooperate to achieve these. Value engineering, widely used in the US, is an outstanding example. Trade contractor and management team work together to find ways of reducing costs or construction times – but not at the expense of quality or performance. For instance, internal staircases were originally intended to be steel frames infilled with concrete on site. Critical appraisal led to them being supplied with concrete already in place. The client benefits from a better, faster job, the trade contractor improves his productivity, the management team increase their expertise.

A vital element in achieving the right atmosphere is using the right form of contract. All too often the terms management contracting and construction management are devalued. Standard forms of building contract are given a change of title when what is needed is a change of attitude. At Broadgate a specially prepared agreement was drawn up to eliminate the barriers, commercial or managerial, between the different members of the team. The results have been spectacular. Fast track combined with fast build have seen phases one and two at Broadgate go up in half the time of 'normal' contracts.

Based on Building, *8 and 15 May 1987, reproduced with permission.*

B.

Housing – cost per bed space
Educational buildings – cost per area based unit of accommodation
Supermarket – cost per storage space
Offices – cost per floor area

The above yardsticks show how designs for different types of building can be assessed quickly and simply for their economic efficiency. Published cost figures can be used as a target for the designer or as a check against tender prices. The government can use them to keep limits on public expenditure and, although there have been complaints that adherence to yardsticks produces unimaginative and low quality buildings, their use can encourage designers and builders to make better use of resources. Of course cost yardsticks must be realistic and must reflect changes in inflation rates and construction methods.

Procurement offers a variety of choices, for example:

- bills of quantities
- lump sum based on specifications
- schedule of rates
- target cost
- performance specification
- design and build
- construction and project management

Each has its advantages. The choice is up to the client who must look for the method best suited to his particular circumstances and building. While open tendering provides for maximum competition, the case against it has been widely aired. Whatever the system that is used, the important thing is that the right contractor is chosen.

The traditional British system has separated the responsibilities for design and construction. This has succeeded in keeping costs down through:

- the use of yardsticks to set limits for architects;
- independent quantity surveyors to ensure architects achieve the required cost targets;
- the pricing skills of estimators who have to apply competitive unit rates to given quantities.

For traditional work the bill of quantities system has proved itself effective. For more complex contracts involving new technology, with the associated problems in managing plant and labour, the traditional system may not be so effective. Hence the growth of alternative methods of procurement.

Based on Building, Technology and Management, *December/January 1986/87, reproduced with permission from the Chartered Institute of Building.*

C.

The noise of heavy construction drowned the more traditional sounds of bat on ball this spring at Lords, the most celebrated of English cricket grounds.

The grandstands are among an increasing number of prestige construction jobs in Britain which are running late, over-budget or are involved in disputes over damages.

Reasons for delays and increased costs include:

- The pace and scale of increased demand in the 1980s which stretched the management resources of contractors.
- Failure to make sufficient investment in training to cope with the upturn when it arrived.
- Over-optimism by firms who promised more than they could deliver in order to win contracts.
- New forms of management contracts used to speed up development. Problems arise when construction starts before design work is completed if alterations then have to be made to the designs.
- The contractual relationships in Britain which creates a climate of fear where contractors are more concerned with protecting their own position than developing a joint approach to problem solving.
- Developers taking advantage of the less buoyant market to insist on small faults being rectified before settling payments.

Richard Griffiths, of Olympia & York, contrasts American and British attitudes. Regular meetings of all groups involved to discuss progress and design changes are the norm in the US, but in the UK 'contractors fear they will face an automatic claim if one set of tradesmen gets to hear of possible delays caused by another group of workers.'

A survey by Johnson Jackson Jeff remarked on the over-optimism in the industry. 'Design teams are so busy that tenders go out without adequate information. Management contractors pass the problem down the subcontracting ladder so that greater strain is placed on the resources of smaller companies.'

In the view of Barry Myers, of Trafalgar House Construction, 'The big difference is that US developers are more closely involved in designing and planning a job; this task is carried out in much greater detail before contracts are awarded ... (which) ensures there are far fewer changes in design during construction.' He feels that the trend in favour of management contracting has failed to generate the same discipline found when a main contractor does the work and takes on the construction risk.

Adapted from an article in Financial Times, *© F.T., 20 July 1990, reproduced with kind permission.*

WORKSHOP

1 Explain the basis of procurement by
- ***a*** bills of firm quantities;
- ***b*** lump sum based on specifications;
- ***c*** schedule of rates.

2*Discuss the economic advantages/disadvantages of open tendering compared with select tendering.
Describe and discuss the importance of the methods used to choose the firms for selective tenders.

3*What were the circumstances which favoured the trend away from traditional procurement towards various forms of management contracting in the 1970s and 1980s?
What are the main differences between management contracting and the more traditional forms of organization?

4 How important is the choice of contract system to the efficiency of the building operations?
How would you advise a client who was uncertain of the best way to proceed with a project for:
- ***a*** the construction of new warehouses;
- ***b*** a city centre leisure complex.

DISCUSSION QUESTION

❑ 'The worst possible situation is to obtain work too cheaply' (contractor's spokesperson). Consider this from the viewpoint of the firm, and of the client.

13 Forms of business organization

PREVIEW

- How are businesses organized?
- Why do some small businesses choose to form private companies rather than partnerships?
- What is the difference between a private limited company and a public limited company?
- Why do some local authorities run their own construction departments?
- Would a state-owned building business offer any advantages not offered by the private sector?

The legal form a business takes is often a function of its size but the owner(s) may take into account a number of other factors, including tax advantages and the extent to which they wish to retain active control of the firm.

SMALLER FIRMS: SOLE PROPRIETORS AND PARTNERSHIPS

There are a great many small firms and self-employed tradesmen in the building industry. These firms have the advantages of flexibility because the proprietor(s) are answerable only to themselves and are able to respond rapidly to changes in the market. Against this their financial resources are often inadequate and the responsibilities of running a business may prove burdensome.

The easiest way of starting up is as a sole trader where one person owns the whole business. Although he or she can borrow funds and hire help, the responsibility, financial and otherwise, rests with the owner alone. If there is more than one owner involved, it becomes a partnership and responsibilities are shared. No formalities are required, though in the latter case a properly drawn deed of partnership is advisable. This sets out the terms on which the partners agreed to share ownership and includes arrangements for terminating the partnership if necessary. (By clarifying such matters at the outset, many potential problems and much heartache can be avoided.) With these simple forms of organization business can start without delay.

Owner-managers in this sort of enterprise have to be jacks-of-all-trades. They make the fundamental decisions about the nature of the business – the type of product or service to be provided, the range of customers to be targeted, the manner in which production is to be undertaken. If the firm is too small to justify hiring office staff, the proprietor has to attend to the paperwork as well as seeing to the job on site. With the growth in legislation affecting business activities, this can be a major headache to small firms. (Frequent complaints are voiced about the amount of unpaid time put in on behalf of the government, especially on filling out VAT returns.) A small builder must decide whether to undertake general maintenance jobs or to concentrate on a particular trade, how wide an area to cover, what equipment is needed, etc. At the same time the owner of a business must find the necessary resources to get it going. If personal savings are insufficient to finance the start-up, the entrepreneur – for that is what the business person is – will have to seek additional funds, either as loans or equity.

Borrowing money imposes a responsibility on the owner to pay interest and ultimately to repay the loan. The loan adds to the firm's expenses. If someone else can be persuaded to take an equity stake in the business, to provide funds on the basis of a share in the profits (or losses!), the business is transformed into a partnership.

The main drawback of these simpler types of organization is unlimited liability, i.e. the owner(s) are personally responsible for all debts incurred in the course of the business. They cannot draw a line between their business affairs and their personal assets. If the business fails, the family jewels and the family car are as much at risk as the office computer and the firm's van.

Liability in a partnership is shared equally, each partner being held responsible for commitments made by the others, with or without consultation. Successful partnerships require mutual trust and sound judgement. Traditionally partnerships have been favoured by architects, surveyors and other professions, where unlimited liability has been regarded as indicating the good faith and personal commitment of the practitioners. Smaller businesses generally have found partnership a convenient form of organization, especially where partners are members of the same family.

LIMITED LIABILITY COMPANIES

Until recently limited liability was only available to limited companies, which ruled out sole traders because a company had to have at least two shareholders. Many traders got round this by setting up private limited companies, with another member of the family holding nominal amounts of shares to qualify for company status. They remained, in reality, one-person businesses. There is now the possibility of limited liability for single shareholder companies. There is no upper limit to the number of shareholders in a company.

The main reason to form a company is to separate business from personal assets. Should the firm collapse, the loss suffered by shareholders cannot exceed their investment. Although this means the risk to shareholders is greatly reduced, the risk of business failure is no less than before. Indeed it may be greater as the entrepreneurs may act more rashly once their personal liability is restricted. In the event of failure, the firm's losses are borne by its customers and creditors, who fail to get the goods they have ordered or the monies they are owed. The advantages of limited liability are that it makes it easier to find investors and it encourages innovation and enterprise.

Many family businesses are organized as limited companies but others prefer the informality of remaining unregistered. A private company does have to be registered and comply with certain regulations, but these are neither expensive nor onerous. Many family businesses have been started on this basis and some have grown to considerable size within this format. The main limitation is that shares cannot be made available to the public, which restricts the company's powers to raise additional capital through new share issues.

A public limited company (PLC) does not suffer from this restriction and it is often the growing business's appetite for funds that motivates its owners into 'going public'. By inviting members of the public to subscribe to the business, it enjoys much wider opportunities to raise funds. Going public also gives existing shareholders greater liquidity as they can now realize the value of their shares by selling them on the open market. Since the shareholders in a family firm may be quite wealthy in terms of the value of their business but at the same time have relatively little spare cash for personal consumption, going public allows them to enjoy a higher lifestyle without losing the business. As long as 51% of shares remain in family hands, they retain outright control. As more shares are sold there is a greater risk of outside shareholders outvoting the original owners and the possibility that they will lose control of the business.

A company dedicated to expansion may seek to go public to facilitate mergers and takeovers. Buying another business is made easier if the company can offer shares in payment rather than having to produce cash to compensate the target company's shareholders. Of course, by the same token, it may become the subject of another firm's growth strategy, as occurred in 1985 when French Kier, one of the top twenty UK construction firms, was snapped up by the much smaller C.H. Beazer. If a company with sound assets finds it is failing to make enough profits to satisfy its shareholders, it will become attractive to would-be buyers. Provided they can convince the majority of shareholders that they will do better with the new owners, the bid will succeed, even though it may be opposed by the firm's directors, perhaps including its founders.

The firm itself benefits from increased status as a PLC. At the same time it is placed under much greater scrutiny, for once it invites the public to put money into it they are entitled to know what it is doing. In addition the costs of going public and seeking a Stock Exchange listing are considerable. For these reasons the public company is a form of organization better suited to larger businesses.

Problems associated with size – diseconomies of scale – were discussed in Chapter 10. Alternative responses were to adopt a regional or divisional structure. The company form of organization allows a high degree of autonomy to these. Each division – for general contracting, design-and-construct, management contracting, etc. – can operate as a separate business with its own organization, its own managing director, estimating department, quantity surveyors and so forth. As wholly owned subsidiaries they report back to the parent company's main board of directors but enjoy considerable freedom in their operational decisions.

Takeovers may follow a similar pattern. The newly acquired subsidiary can continue to operate as an independent unit, even retaining its own name. Alternatively it may be absorbed into the parent company, losing any separate identity. There is no ideal structure – circumstances and fashions change. At one time centralization is praised for its economies of scale, with shared services such as estimating, purchasing or design. Another time the emphasis shifts to integrated teams and local autonomy. Constant change is unsettling, but periodic reviews can stimulate new ideas as well as recognize changes in workload and in personnel.

Large-scale construction work, such as the Channel Tunnel or the regeneration of London's Docklands, goes beyond the capability of a single firm to supply. A consortium, or grouping of firms to undertake the particular project, provides a solution. This is becoming more common, especially in international markets, and is likely to get a further boost as Europe becomes more closely integrated. Cooperative societies of the traditional sort, owned and operated by their members, are not significant in construction. Self-build associations with the limited aim of providing housing for their own membership have gained some popularity, but they are temporary organizations and have not much impact on the wider market. They may play a larger role in the future, as the government places more responsibility for social housing with housing associations. In some areas this has encouraged shared equity schemes, whereby houses are part owned, part rented. A number of self-build groups have been set up on this basis, but they remain on the fringe of the market.

PUBLIC SECTOR CONSTRUCTION ENTERPRISES

State ownership is also limited, being confined to the direct labour organizations (DLOs) of public authorities. A more broadly based public sector involvement has been proposed in the past, notably in the Labour party's document *Building Britain's Future* (1977). This suggested setting up a National Construction Corporation on similar lines to other public sector corporations. The management structure of such organizations is broadly similar to that of a public company, but the directors are accountable to Parliament and must follow the policy guidelines laid down by the government, whereas a company is answerable to its shareholders.

At present, with widespread privatization of public enterprises, nationalization appears to be a dead issue. Nonetheless it is worth looking at the arguments put forward in its favour. The case for public sector involvement in construction rested on two broad planks. Firstly there is the nature of the product. Housing, a major part of the industry's output, is seen by many as a merit good, one that is socially desirable but where the cost often exceeds the consumer's ability or willingness to pay. In addition to housing, the construction industry provides a large part of the nation's infrastructure, in the form of road systems, power stations, etc. These are vital to the nation's welfare and the nation, through Parliament, should therefore have a direct voice in the organization of the industry which provides these goods.

The second plank in the nationalization platform concerns the working practices of the industry. Despite many – perhaps too many – regulations it has a poor safety record and remains heavily reliant on casual labour. It suffers from the image (and practices) of the cowboy operator. Greater job security, improved working conditions and better quality control may be more easily attained where the terms of contracts, standards and training are matters of policy and not determined by profits. A public sector construction corporation was seen by its advocates as a trail-blazer organization which could set standards for the whole industry.

Nationalization, in the manner proposed, did not try to establish control of the whole industry, but only aimed at providing a public sector alternative to private enterprise. This is not unlike the current position of local authority direct works departments who have to tender for local authority work in competition with outside contractors. Many of the arguments about the role of these direct labour organizations (DLOs) mirror those about the extension of public ownership. Today the climate of opinion favours relying on market forces with separate watch-dog bodies to regulate matters which fall outside the market and protect consumer interests in the face of monopoly suppliers.

The emphasis on markets is bringing DLOs more into line with the private sector practices. Since the first DLO was set up by London County Council in 1892, they have had fluctuating fortunes. Other authorities were slow to adopt the system, but by the Second World War it was well established and in the crisis years after the war DLOs reached their peak. Thousands of new dwellings were erected by DLOs in the 1940s. In the 1950s they began to decline and today their role is largely confined to repair and maintenance work.

DLOs are often castigated for their inefficiency, but the evidence is scanty. Opponents quote lower productivity figures, but fail to take into account the differences in the type of work undertaken, which tends to be biased towards minor repairs and is therefore more labour intensive. Other costs and benefits should also be considered, for instance the generally higher standards of safety and provision of training. At the same time public accountability can lead to a high level of bureaucracy, which may be seen as wasting resources.

Since 1981 DLOs have been required to show a return on capital employed. Unlike private sector firms, they had to conform to strict accounting procedures which did not permit losses in one category of work to be offset against 'profits' in another category. A substantial proportion of larger jobs is put out to general tender, but whilst outside contractors can compete for these jobs, DLOs were restricted to public sector jobs. It has been argued that this is requiring DLOs to compete with their hands tied, by imposing conditions more onerous than those which operate in the private sector. Since private and public sector organizations differ so much in their aims, workload and structure the debate is likely to generate more heat than light.

Increasing numbers of DLOs have been turned into independent enterprises so as to compete for outside work. Where this has been done without changes in the management, lack of commercial experience and skills have led to substantial losses in some instances. Such cases are quoted by their opponents as evidence that the DLOs were inefficient, but it could be argued that it merely shows they are trying to undertake a new role, in which they are not skilled, without proper training and guidance. What is needed is a clearer view of the functions of DLOs and how they are best performed.

CASE STUDIES

A.

In 1972, Sir Robert McAlpine, a family business for over a century, finally decided to go public. The decision was not taken lightly, for it meant an important change of style for the company. Risk takes on a new meaning when the interests of outside shareholders have to be taken into account. For instance, Pilkington Bros, who took the plunge a year before McAlpine, would likely have hesitated over spending so many millions of pounds in developing float glass manufacture if outside shareholders' interests had prevailed at the time.

The McAlpines had always avoided publicity, but going public meant issuing a prospectus detailing the firm's track record, its directors, their plans and prospects. Every little change of fortune concerning contracts, turnover, disputes, whether fact or fiction, is a matter of public interest when share prices are involved.

At the time, the business was estimated to be worth £20 to £30 m, which meant several of the family became instant millionaires. Being able to sell shares made it possible for them to raise cash without causing a headache to the rest of the family. Ready cash may mean luxuries, but it becomes essential when death brings tax demands. Pilkington's earlier flotation had been precipitated by such needs after three deaths in a single year.

Outside investors provided opportunities for expanding the business which even McAlpines, a firm which had always managed to generate sufficient funds for its needs, could not ignore. Construction in the 1970s was a rapidly changing industry. Bigger contracts and new technologies had been developed. Going public was a vote for the future.

B.

Government policies to promote competitive markets extended deregulation to chartered surveyors and estate agents in 1986. Up to this time RICS rules had limited them to operating as partnerships; the opportunity to incorporate their businesses meant a number of factors had to be considered.

Former partners who set up a company change their status into employees of the business, with implications for tax and pension provisions. As partners they share in profits, and taxes on profits are not usually due until the year following. With salaried staff, tax is paid on income as it is earned, but against this a better allowance is made for pension contributions. Costs of indemnity insurance also have to be considered; limited liability would help to keep down the premiums as the firm cannot be held responsible for debts exceeding its assets. Naturally claims for negligence or fraud can still be brought.

As a company the business would have to hold annual general meetings of its shareholders, its books would be subject to an annual audit and its returns could be examined by interested parties at Companies House. All this means a certain loss of intimacy, of informal and flexible procedures. Similar difficulties could arise in passing on the practice to younger members. A relatively junior shareholder might be unable to buy out a senior who wished to retire; nor would it be easy, in a private company, to establish a fair price for the shares. This is especially so where the firm's main assets are its reputation and expertise.

For smaller firms the chance to form a company was not an overwhelming advantage. Their personal style of business remained well suited to partnerships. Bigger firms, who were already finding the structure of a partnership inhibiting to the more streamlined management needed by a large business, gladly took the opportunities of forming companies. This gave them access to outside finance and the limited liability needed to protect these shareholders. It also created new possibilities for takeovers and the emergence of chains of property agents with branches nationwide, competing with the more traditional firms.

c.

Bill Brown is angry. 'I get fed up being told I'm not productive,' he said. 'Local authorities contribute as much to the local economy as anyone.' His appointment to head Lambeth Council's construction services comes in the wake of a damning report into the department.

The investigation followed a series of internal audits, arrests over alleged corruption and suspension of two contractors from the tender list. There was particular criticism of the management which had been reluctant to right shortcomings. Recommendations include immediate adoption of a set of financial regulations and a revision of tender lists to ensure competition. Last year the DLO, which employs 1400, reported losses of £1.8 m on expenditure of £37 m. Most of the losses were incurred on work over £50,000.

Statistics produced by the Chartered Institute of Public Finance show DLOs generally are good at road repairs and general maintenance, less good at new build, in particular larger projects like housebuilding.

Brown reckons there are things which DLOs do better than other companies. He points out that when it comes to performance targets, profit is not necessarily a good indicator of quality. 'One of the essentials of a DLO is the terms and conditions of its workforce. We must be efficient but that can't mean cutting wages and conditions.' He points to Lambeth's achievements in recruiting from minority groups and its training record. 'Local authorities train more than their fair share of apprentices. Not providing a pool of skilled labour is a crazy thing for the construction industry to do.' He also remains staunchly in favour of contract compliance on issues like health and safety.

Brown remains uncompromising in his defence of DLOs which he sees as a skill resource. 'They are the only organisations really looking at the long-term needs of the industry. There are good private firms and we work very successfully with subcontractors, particularly local builders. But a DLO's links with its local community is a strength that most firms do not have.'

Adapted from Building, 9 October 1987, reproduced with permission.

WORKSHOP

1 *What are the key differences between:
  a a partnership;
  b a private company;
  c a public limited company?
  Explain the significance of these differences.

2 Why should a successful private company consider a stock market flotation? Why might a PLC wish to buy out shareholders and revert to private company status?

3 Explain the benefits that could result from setting up a nationalized housebuilding corporation.

4 Is profitability, or return on capital, a good way of assessing the efficiency of a DLO?

DISCUSSION QUESTIONS

- Should the criteria for judging the success of an enterprise differ for public sector and private sector organizations?
- 'The quality of the entrepreneur is central to the success of the enterprise.' 'A strong team is key to success.' Can these views be reconciled?

14 Marketing

PREVIEW

- Is marketing the same thing as advertising?
- What can a contractor learn from market research?
- Do builders have anything to gain from marketing their services when most clients have decided what they need before they contact a builder?
- What is corporate image building?
- What does marketing have to do with tendering?
- Do small firms need to bother with marketing?

LOOKING AT MARKETING

The construction industry often regards marketing as an irrelevance. In the 1960s, when demand exceeded supply, contractors had little need to worry about customers. Moreover, the central concerns of marketing experts, encapsulated as the four Ps – **product, price, promotion** and **place** – were not, except for price, the contractor's primary concerns. The product was generally determined by the client or architect. Place was decided by the client and there was little point in promoting a service that the customer had already decided to buy. Small builders who mostly dealt in maintenance work did not even have to seek out tender opportunities. Instead their customers approached them.

The more turbulent climate since the 1980s has created greater awareness of the benefits of strategic business planning. Looking to the firm's future raises questions about its client base, about market trends, about the firm's own structure and its resources. Forecasts have to take into account wider economic and social changes, such as demographic trends, as well as specific changes like the introduction of new building standards. Marketing is no longer on the sidelines.

What is marketing? Too often it is still looked on as yet another cost, one which does not contribute directly to output. Is it just another word for advertising or selling? Market research is associated with clipboards and questionnaires. None of this seems particularly useful to the contractor.

The concept of marketing is both simpler and more complex than this amalgam of advertising, selling and questionnaires. Marketing means matching the needs of the customer more closely with the services and products which the firm can supply. It is interactive. It is about customer satisfaction, but it is also about the bottom line – producing at a profit. This puts marketing at the heart of management (Fig. 14.1), as the function which informs all other aspects of the business: the choice of contracts, bidding strategy, the use of labour and organization of plant, how to raise the finance, the design of products, specifications and the quality of jobs, even after-sales service – in short everything in the business that is directed towards customer satisfaction and consequent profitability.

LOOKING AT CUSTOMERS

Research is the starting point. Customers' needs and wants must be understood, especially in the competitive market for speculative building. There is little advantage in building homes with extra bathrooms if buyers really want an extra garage. Identifying market trends and opportunities helps a

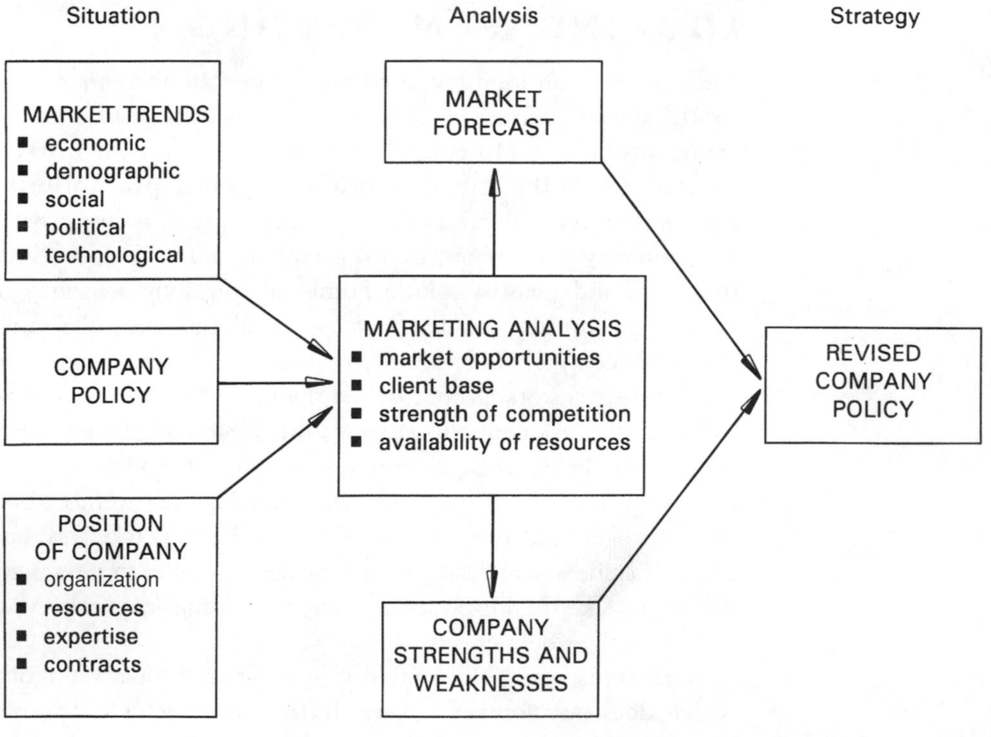

Figure 14.1
The functioning of marketing.

business to find customers. Builders who saw the significance of changes in the age structure of the population introduced a new concept in housing – sheltered housing for the retired. (It could be argued this is nothing new – almshouses for the aged were being founded in the reign of the first Queen Elizabeth and earlier, but the concepts are not quite the same!) Firms which spotted the opportunity early and researched the needs of this age group have prospered by filling a gap in the market.

As well as looking for gaps in the market, the firm needs to identify what it is that the customer is buying. This is not always as obvious as it seems. A compact, well-fitted bathroom may be perfect for washing and shaving, but can it cope with bathtime-cum-playtime for the kids – is there space for lots of towels and toys and cast off clothing? Is the kitchen seen as a functional workroom or the centre of family life? People rarely look for a single quality when choosing a purchase; they look for a package of benefits – performance, appearance, convenience, price, etc. Successful marketing involves providing a product or service which overall gives the buyer a better package and hence more satisfaction (**utility**) than can be found elsewhere.

Where does a construction company begin with the important task of researching the market for new openings? A good example of the sources and type of information used by one company is the way Tarmac approached this problem in the 1970s when they realized that their main activity, road building, was slowing down. After a period of growth motorway networks in the UK and Europe were no longer expanding. Tarmac needed to identify new opportunities. They decided to look overseas for areas where they could use their expertise in roads. They began with official, published statistics. (Governments provide huge quantities of information of social and economic trends.) They compared countries in terms of their area, the size of their existing road networks, their population, income and national output. These statistics helped to identify a number of countries with relatively sparse road networks, an adequate level of population and a high rate of economic growth. The combination of rising incomes with a limited existing road system provided a pointer to growing demand in the near future for more transport and new roads.

Further research reduced the list of possibles to a smaller group of probables. At this stage attention turned to factors such as the political stability of each country, its capacity to mobilize funds through banking systems, skills available in the indigenous labour force, etc. When the choice was narrowed down to half a dozen areas, even more detailed feasibility studies helped to determine where the first approaches should be made. People with recent experience of working in each country were asked for their assessments. They came from different backgrounds – diplomatic, financial and commercial, including other UK companies already in the area. Finally an on-the-spot assessment was made of contract opportunities, resources and potential partners within the locality. The careful preparation paid off and Tarmac penetrated new markets overseas.

For a small business, research is much less formal. If a firm keeps adequate records of its jobs, it has an internal source of data as to which customers, and what type of jobs, have provided the bulk of the workload and which have contributed most to profits – not necessarily the same thing. An informal chat with a customer, especially if it occurs as a follow-up soon after the completion of a job, can ascertain the level of satisfaction and any causes of dissatisfaction. It is a natural tendency to forget the jobs which turned sour, but knowing what went wrong helps to avoid a repetition. Marketing is about identifying weaknesses as well as building on strengths.

LOOKING AT OTHERS

A firm needs to make comparisons with others to assess the competitiveness of its products, services and prices. Some firms make a point of recording such data 'because we must know how well we are doing in relation to the total market'. Even in small firms, managers need to keep their finger on the pulse of market activity. Employees and customers may have experiences of dealings with other firms which can help to sharpen perceptions of how the market is moving and keep estimates from becoming consistently too high or too low. Preparing bids for tender, deciding whether or not to compete for a particular contract, should be seen in the marketing context.

Bidding decisions tend to be opportunistic, guided by what is available and current workloads. If there are too many competitors and little likelihood of winning the job, or doubts about the successful completion of the work, a firm may prefer not to bid. Sometimes they openly withdraw from the competition, but if they fear this could lead to loss of invitations to bid in the future they are more likely to put in a price which is intended to be realistic, but not to secure the contract. A 'cover' price can be organized by consulting a friendly rival.

Pricing policies reflect marketing strategies. Firms which compete aggressively on price risk losing their profit margins if problems arise on the job. Often they aim to compensate by pushing up prices for any variations to the contract. Other firms prefer to build a reputation for controlling costs by avoiding very low bids and keeping prices down on any extra work.

LOOKING INWARD

The traditional approach, via competitive tendering, places the firm in a rather passive role, reacting to opportunities as they arise. Marketing oriented companies prefer a more proactive stance, deciding what type of work they want and looking for suitable contracts. This approach may take them away from traditional procurement towards alternatives which give the contractor more freedom of action. Design-and-build, fee management and negotiated contracts all give the firm more scope to differentiate its services and build an expertise to attract clients.

The size and diversity of the construction industry makes it difficult to serve all needs; some sort of **market segmentation** is needed to focus attention. Construction markets can be divided in all sorts of ways. Broad categories might be based on types of work – infrastructure projects, commercial work, industrial buildings or housing — or on the size and value of contracts, or the geographical area to be served. Within these limits further segmentation can narrow the field as much, or as little, as the company feels appropriate. Several areas may be identified as suitable markets so the firm must decide whether to opt for diversity and spread risks, or develop expertise in a particular market.

Market research involves looking inwards as well as looking outward. Evaluating the firm's position in the market, its strengths and weaknesses, is part of making sure it has the resources to meet its objectives. The old motto 'Know Thyself' sums up the importance of being aware of the unique blend of talents and weaknesses which make up any one firm. A **marketing audit** is a process of collating data to assess the firm's current position and review its policies. Even the smallest firm should periodically review its position. Of course the process would be much less formalized for a small business. A general builder might draw up a balance sheet which included:

- Market situation:
  - – slow housing market;
  - – high interest rates;
  - – repair work steadier than new building;
  - – cutbacks in local authority jobs;
  - – more competition, more fringe operators around.
- Strengths:
  - – good quality of workmanship in general;
  - – good reputation: many customers return;
  - – new customers recommended by existing customers;
  - – good relations with suppliers;
  - – reliable labour force.

- Weaknesses:
  - several complaints about plumbing jobs;
  - records not up to date, inadequate for this audit;
  - too many customers late in paying;
  - workload too variable (seasonal);
  - dependent on very localized area.

This exercise helps the firm select suitable jobs and turn down unsuitable ones. It reveals positive features of the business which can be emphasized in advertising the firm's services. They may also form a basis for growth. Action to remedy weaknesses might include hiring or training labour to fill skill gaps, seeking suitable subcontractors with whom to build a relationship, investing in the right equipment, etc. Alternatively the firm can withdraw from areas where it is not successful. Marketing weaknesses that might prove difficult are the seasonal and local nature of its workload. Developing services to complement its current activity and targeting a specific neighbourhood could be considered.

Haphazard growth can result in a firm becoming over extended and losing control over its workload. Smaller firms in particular often feel a pressure to take on whatever jobs come their way, but with limited resources this is a risky tactic. The firm with a defined market strategy, reviewed often enough to retain flexibility, has more control over its future. Many firms who were building for first-time buyers in the mid 1980s saw no need for review – business was booming. But even without recession, marketing minded firms might have been planning to develop other sectors. With two-thirds of householders already owner-occupiers and the proportion of under 25s in the population falling, the market for first-time house buyers was likely to slow down. Firms which specialized in this type of work needed to reappraise their strategy accordingly.

LOOKING AHEAD

Product development and innovation are part of the drive to provide clients with a higher level of satisfaction. If the idea of product development seems inappropriate to, for example, a housebuilder, the reader should review his or her understanding of the 'product'. A house is not simply a house; it is a package which encompasses space for various activities, such as sleeping, cooking, entertaining, watching TV, studying, etc. It provides shelter from damp, cold and heat, as well as providing its own environment with visual, auditory and other aspects. Its location influences access to jobs and social facilities. All of these elements are weighed up by the potential purchaser. Today's housebuyers have different aspirations and expectations to earlier generations – changes which need to be recognized.

As well as researching the customer's wants marketing serves to modify them. Advertising and promotions provide opportunities not only to sell, but to inform. Understanding the limitations of a product as well as its strengths, and the technical and economic constraints on production, can help buyers to abandon unrealistic expectations. A simple example might be advice on proper ways to clean plastic finishes. The householder is saved the disappointment of surfaces being dulled by abrasive cleaners and the builder avoids being blamed for supplying poor quality goods. After-sales service and follow-up research continue the dialogue between customer and supplier, providing feedback for the ongoing process of development and improvement.

Not all promotions centre on a product or service. Some are aimed at raising the public's awareness of the firm and enhancing their view of it. The direct impact on sales is difficult to measure but until Cornhill Insurance sponsored test cricket few people could have named the company or its business – now it is known by millions. Such corporate image building is mostly undertaken by larger companies, but within a local context there are opportunities for small firms to make more use of sponsorship schemes and similar promotions to raise the awareness and goodwill of the community towards them. Logos and symbols are aids to public recognition, creating a sense of familiarity and trust. At the same time a well chosen device conveys an appropriate message about the firm – think of the Barratt oak tree (strength, tradition, stability) or the Wimpey cat welcoming buyers to a cosy, comfortable home.

By now it should be clear that marketing is not a separate function that can be dealt with apart from other aspects of a business. It permeates all activities which bear upon the customer's ultimate satisfaction and the public's perception of the firm. Every decision about what to produce, how to produce it, every contact with people outside the firm, whether suppliers, competitors, customers or bystanders, all have some impact on how people see the firm. An active marketing policy puts these issues on the agenda instead of allowing them to go by default. Marketing is at the heart of strategic planning in the boardroom.

Marketing also operates at the tactical level, in the details of daily business. Messy sites, dirty transport, badly written correspondence, all have a negative impact on the public's view of a company. A friendly and efficient receptionist, prompt attention to enquiries, organized sites, well maintained plant and a host of other details contribute to a positive image. Public perceptions matter because the public are potential customers. They are also potential suppliers and potential employees, all of whom are part of the successful business equation.

The construction industry prides itself on a tough, down-to-earth approach to providing practical solutions to real problems. Yet anyone who looks at the high failure rate of firms in the industry must realize that one very real problem is survival. The self-employed tradesman has as much to gain from marketing as the international company. The methods will differ, but the aims are alike – a profitable business with a high level of customer satisfaction. In a competitive, commercial world marketing skills are an essential aid to survival.

CASE STUDIES

A.

In the mid-1960s Charles and Susanna Church sold their home in Camberley and moved into a hut to finance their first new-built houses. Customer contact then was a matter of a trip down the road to chat with new owners to see if everything was alright. As the business grew, a more formal system evolved, where quality control and after-sales form a circle of information and response that feeds owners' experiences of Church's house designs into the forward planning process.

'Taking to people,' says Susanna Church, 'is the best way of seeing how individual houses behave, so we call on people some months after they have moved in and ask them how they feel about the house. That leads to design alterations in later houses, details such as the way doors are hung, where the TV points are, positioning of electric plugs and so forth as well as specifications for the houses.'

As the business has grown, so has the after-sales department. A permanent maintenance team has been added but homebuyers' calls cover far more than the usual crop of teething problems. Landscaping and problems with the neighbours are part of the job. In her time handling calls from new owners, Susanna Church found that 'the people who do run into problems seem to be those who, months after they have moved into a house, are still disorganized. I remember one call from people who had found a single screw loose in a door hinge quite a time after they moved in. Obviously that's something that shouldn't happen, but, really, you would expect that most people would have got a screwdriver out and tightened it themselves.

Making sure the screw is tight in the first place would be the responsibility of the other, quality control, side of the operation. Marion Brewer took over that job in the late 1970s. Since then her work has moved further down the planning process, so that it now includes selecting the type of brick and tile finishes, as well as choosing and ordering all the fittings, from kitchens to built-in wardrobes.

When the builders have finished, Marion Brewer's quality controllers go around completed houses with a checklist to see everything fits.

Knowing what people want calls for a combination of experience and, according to Susanna Church, imagination. 'You have to imagine you're living in a house to get a feel for it. When we have a new house type and we are setting up a completely fitted showhouse, you have to imagine the type of family it might appeal to.' In one case this was an airline pilot, with a wife who has a part-time job and a couple of young teenage children.

It is a far cry from the 'build it, sell it, move on' days of housebuilding. But even with the most detailed market research every new house style is, to an extent, a gamble. As Susanna Church says, 'We'd love to be able to build an ultra-modern house that people would buy, but I've never seen one that would sell well. Traditional homes are what people want.'

Extract from Financial Times, © F.T.,
16 January 1988,
reproduced with kind permission.

B. Marketing audit: checklist for review

External factors

- Economic environment:
  - state of economy
  - government policies
  - social climate
  - financial environment
  - changes in technology.
- Market environment:
  - growth/decline in size of market
  - changes in tastes
  - product innovation
  - price trends.
- Competitive environment:
  - major competitors
  - market share
  - diversification/concentration
  - entry barriers/opportunities
  - profitability.

Internal factors

- Company strengths and weaknesses:
  - sales breakdown
  - profits analysis
  - marketing objectives
  - marketing procedures
  - research
  - promotion
  - feedback.

WORKSHOP

1 Explain briefly each of the following aspects of marketing and show how they can be employed in the context of housebuilding (see case study **A**):
   a market research;
   b market segmentation;
   c product development;
   d promotion;
   e after-sales service.

2 Consider how a civil engineering company, with expertise in hydroelectric projects, might apply the above concepts.

3 Distinguish between marketing, promotion and advertising.

4 Refer to the checklist of items for a marketing audit (case study **B**). Give an example of how a speculative housebuilding company might use such an audit, indicating the sort of information appropriate to each heading.

5 Examine the advertisements in a recent issue of a trade journal. Choose one that you find effective and one that fails to hold your attention. Analyse the reasons for their differences in impact.

6*What do you understand by 'marketing'? How important is marketing in the construction industry?

DISCUSSION QUESTION

❑ Do construction firms need to develop their corporate image? If so, what sort of qualities should they seek to project and how can they do this? Relate your discussion to a firm known to you.

15 Productivity

PREVIEW

- What does productivity measure?
- Why is productivity important?
- Does higher productivity mean lower costs?
- How does construction productivity compare with other industries?
- How does construction productivity in the UK compare with other countries?
- How can the industry improve productivity?

PRODUCTIVITY AND EFFICIENCY

Productivity is concerned with the ratio of output to inputs, singly or in combination. Most often it is measured in terms of a single resource, i.e. labour. Strictly speaking this should be called **labour productivity** and it is generally quoted as a ratio of output per man or man-hour. Total factor productivity measures output against the combined inputs of labour, land and capital, suitably weighted and expressed in money terms.

Thus if two workers are employed to excavate trenches using pickaxe and shovels and they remove ten cubic metres of soil in an eight-hour day, their productivity is 0.625 cu.m/man-hour. If a third man joins them and output rises to 15 cubic metres, productivity is unchanged. If they are now given a mechanical digger to use and output rises to 30 cubic metres, labour productivity has doubled to 1.25 cu.m/man-hour. However additional resources have been employed in the form of capital equipment, so total factor productivity has not risen to the same extent. Quantifying this will depend on how labour and capital are aggregated. If the digger is valued at the same rate as two labourers, the inputs have risen to the equivalent of five workers and total factor productivity is 0.75 cu.m/combined factor-hour. Calculating relative costs of workers and machines poses problems, hence the use of simpler, if less revealing, labour productivity measures.

Productivity does not necessarily mean efficiency. The latter is a far broader notion, concerning the effective use of resources to achieve the desired outcome. There are possible conflicts of interest here, e.g. between client and contractor or contractor and subcontractor. Efficiency for the contractor may mean giving priority to another job, but for the client it means getting this contract completed. Efficiency in using resources must recognize their opportunity costs. For instance, heavy investment in plant will nearly always raise output per man but it is not always efficient. If the mechanical digger in the earlier example cost more than hiring three extra labourers, it would not be an efficient way of doing the job. In less developed countries, such as Nepal, the sight of people shifting roadstone in baskets on their heads may appear inefficient, but it is not. Capital is scarce in Nepal, equipment has to be imported and that makes it expensive, but there is a large pool of labour which costs little to employ. So shifting roadstone by hand may mean lower output per man-hour, but it is cost-effective.

Because construction output is not standardized it is difficult to measure. Buildings cannot simply be counted, still less can we add up the number of repair jobs. Output can only be aggregated in monetary terms. Productivity is taken as the value of work done divided by the size of the labour force. Statistics come from Department of Employment, the Inland Revenue, from the census of production, etc. Gross output includes the costs of materials. Deducting these gives net output which shows the value added per man. Gross figures are easier to obtain, but can be misleading because changes in the price of materials, without any alteration in the labour content of a job, can give the appearance of changes in productivity.

USEFULNESS OF PRODUCTIVITY MEASURES

If productivity is so tricky to measure what is the purpose of doing so? Knowing that output per man is worth £*x* p.a., by itself, is little use. It is only in comparison with a previous year, or another industry, that it becomes helpful. Productivity figures can be used within a company to compare progress on different sites. Variations may be due to reasons beyond the company's control, like different physical conditions, the behaviour of the client or the nature of the design, but low productivity may be a warning that action is needed – perhaps that the site is badly managed or the operatives are not properly skilled. Comparisons over time indicate the rate of progress within a firm or an industry and comparisons between industries monitor shifts in the structure of the whole economy. International productivity figures help to assess our competitiveness in world markets.

Valid comparisons can only be made where statistics are compiled on the same basis. Even measurements which are made in the same way can conceal differences which are not strictly related to physical output. We noted in Chapter 13 criticisms commonly made of low productivity by local authority DLOs. On average public sector pay is lower and fewer hours of overtime, paid at higher rates, are worked. If output is measured in money terms and the value of the work reflects the labour cost, these factors alone will depress DLO productivity figures.

Despite these problems, statistical evidence does show the construction industry to be low down the labour productivity league table. Comparison with other countries reinforces this picture. Statistics drawn from official sources have been supplemented by studies of comparable projects across national boundaries. These have ranged from the construction of power stations and other major civil engineering contracts to factory and house-building. Comparisons on various criteria – the total time taken, cost per square metre of floor area, output per man-hour on specific tasks such as pipe-laying, etc. – have all tended to confirm the general conclusion that in construction, as in many other industries, the UK is behind its European and American counterparts, although it has been catching up in recent years (Fig. 15.1).

As a nation our lower productivity underlies low growth rates and slow improvements in living standards. Unless we raise productivity we cannot raise incomes without running into the danger of inflation, which leads to further loss of competitiveness in world markets. To enjoy more consumption we have to increase our production. Productivity is a major issue for the UK economy as a whole and for construction in particular. The reasons for low productivity in the industry are partly within its control – in the structure of the industry and the way it organizes its resources. Other reasons lie beyond its control in the nature of construction work (see Chapter 11).

SITE PRODUCTIVITY

There is no such thing as a problem-free project in construction. Every site is unique and previous experience will have to be adapted to new situations. Scope for improving productivity depends on how far problems are anticipated and avoided. The ideal situation would be an experienced client, a design which gives priority to ease of building, specifications complete and correct in every detail, the right resources available in the right combinations at the right times, a totally efficient management and workforce, plus favourable ground and weather conditions!

In reality some delay is inevitable, but productivity benefits when these

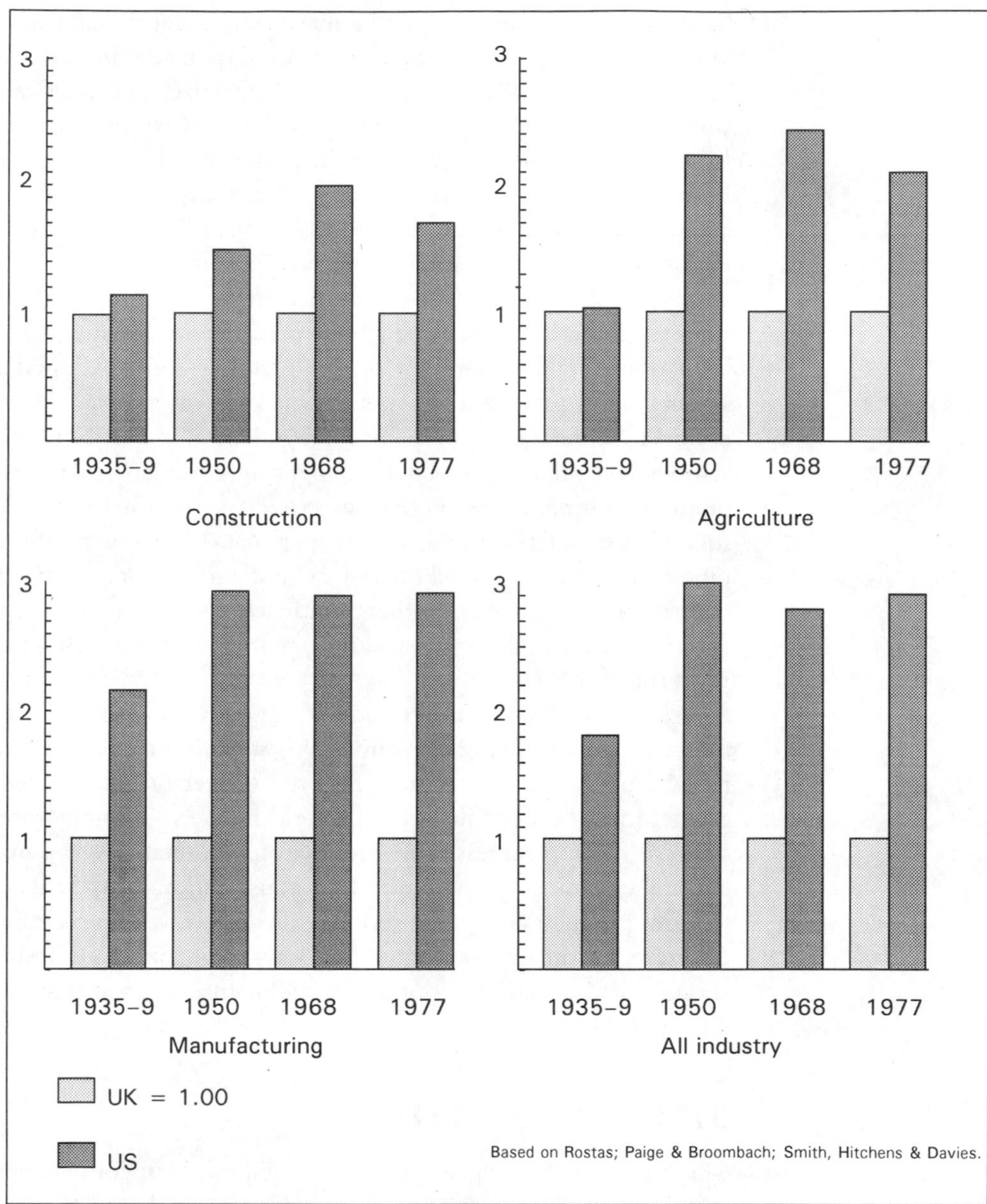

Figure 15.1
Comparison of UK and US productivity. (Source: National Institute for Economic Research, 1982.)

are kept to a minimum. External problems range from the unsurprising, like bad weather, to the unexpected. Work schedules and site facilities should take into account likely weather conditions, while contingency plans are needed for emergencies. 'Unforeseen' ground problems frequently cause hold-ups, but could often be avoided by proper site investigation. Anticipation and planning are the keys to keeping work on schedule.

Some problems lie inside the industry. Poor design and inadequate specifications are a major source of additional work. Sometimes inexperienced clients or a poor brief is at fault. A National Economic Development Office (NEDO) report, *Faster Building for Industry* (1983), found that 45% of the projects studied suffered some delay due to variations introduced by the client. However, not all design difficulties are due to the client and while designers cannot be expected to find the perfect solution to every need, better design could eliminate some of the extra workload. Lack of standardization, excessively strong components, awkwardly placed services, specifying hard-to-obtain items – the potential for greater efficiency is vast. Many of the difficulties spring from the divisions between construction and design, something which needs to be tackled at industry as well as individual project level.

One of the problems frequently encountered on site is the demarcation between trades. If a door or window sticks, it is a job for a carpenter, not a painter, but if the painting has already been done and gets damaged when the carpenter eases the door, then it is not his or her job to make good the paintwork – the painter must be recalled. The carpenter is anxious to get on, especially where the job has been subcontracted and a price agreed for the work. More flexible working practices can save time, raise productivity and reduce costs.

Productivity on site reflects the quality of management. A badly arranged site wastes time in excessive fetching and carrying; poor choice of equipment increases the time and effort to do the job; inadequate supervision leads to faulty work. If materials are not delivered on time, or plant is not available when needed, productivity suffers. The right combination of resources is equally important. If there are too many lorries to remove spoil, they waste time queueing to be loaded; if there are too few, labourers will waste time waiting to remove the debris. The NEDO survey highlighted subcontracting as the most frequent reason for hold-ups, partly because subcontractors had difficulty in rescheduling their own programmes when delays occurred.

Construction managers seek to maintain a continuity in the workload; keeping plant and labour employed is more efficient than lurching between periods of intense activity and slackness. Too much emphasis on productivity can push up costs, by sacrificing continuity to speed of completion. Very tight schedules which involve a series of operations on a site, each undertaken by a different group of operatives moving round the site, run the risk of bottlenecks developing if there is too little headway between tasks. Work is speeded up at a cost of paying for idle time while one gang is waiting for another to complete its task. Having said this, cost over-runs are more likely on projects that finish late than ones which are completed quickly.

Overall the general level of morale will be reflected in the speed and effectiveness of the work. Slow working can damage productivity as much as spells of waiting time. Excessive speed can cause errors which then have to be rectified. Cutting corners also increases the risk of accidents. The industry has a poor safety record, which damages morale and incurs costs, both to the company and to society.

PRODUCTIVITY AT INDUSTRY LEVEL

The cyclical pattern of demand makes it difficult to keep workers employed on a continuous basis. Much of the slack is taken up by using casual labour and subcontractors, but firms are often reluctant to shed directly employed staff. A downturn in demand thus leads to falling productivity. When there is not enough work to go round the rate of production tends to slow down; workers are not eager to work themselves out of a job. The upswing produces different problems as firms struggle to cope with growing work-loads. Skilled operatives may be hard to find – bottlenecks and delays ensue. In these circumstances productivity is slow to pick up.

Because labour supply is more easily increased or cut back than capital there is a tendency to respond to demand by taking on more labour rather than by increasing investment. Investment in the industry is relatively low, a problem exacerbated by the numbers of small firms, many of which are under-capitalized. Small firms cannot afford a high level of investment. They are not innovators and tend to rely on traditional, labour-intensive methods. Less plant and machinery leads to lower labour productivity.

UK consumers also have a strong preference for individual houses of traditional construction. Apartment blocks and industrialized systems may improve productivity but they are not popular. Even in city environments, where buildings of grand proportions are familiar and even admired, local residents have still expressed their preference for low-rise housing in traditional styles with garden surrounds. Architects may not approve of important sites being used in this way – one such housing project, the Coin Street development on London's South Bank, drew the comment in *Building Design* magazine that 'the scale feels all wrong' – but the local authority here allowed householders to control the outcome. (This opens up a much wider debate about the role of architecture which cannot be continued here.)

Architects are party to the traditional separation of design and construction, itself a source of problems. Responsibility for a project generally rests with the architect but communications between designer and builder are weakened by the very different training and experience that each has. Failure to incorporate building expertise at the design stage produces problems which then have to be solved on site instead of on the drawing board. Ease of building – or 'buildability' – is not always the architect's first priority. Delays occur even before work on site begins. The initial time taken to obtain planning consents can be considerable, particularly if there are objections to be heard. The go-ahead, when it comes, may be subject to qualifications which require changes in the specifications and yet more delay.

Some of the problems are inherent in the nature of the industry. Repair and maintenance work is largely unsuited to mass production methods. Where handwork is needed productivity is likely to remain low. But these features are common to all construction industries; the poor performance of the UK industry suggests that at least some of its problems are solvable. Indeed, considerable progress has been made over the past decade. We must look to the organization of the industry and beyond – to its management, to its workforce, to the social and political environment in which it operates – for our answers.

CASE STUDIES

(See also case studies in Chapter 12.)

A.

A report published by the National Economic Development Office (NEDO) on the construction of two near identical plants for Eastman Kodak, one in the UK, the other in the US, reveals disturbing differences in productivity, estimated by NEDO to be 42% higher in the US operation. Both contracts were managed by Bechtel and were under construction at the same time, although an earlier start in the UK meant a slight advantage for the US team who were able to avoid some problems picked up on the UK site.

Building regulations and working practices accounted for part of the differences in output, e.g. the greater amount of scaffolding required in the UK, but the main feature on which the report remarks was the greater level of activity on the US site. British workers applied themselves well when on the job, but time lost on tea breaks, on starting and finishing, meant an estimated 30% less time was spent on production compared to the Americans. If bad weather meant loss of working on the US site, it was made up during rest days, without extra payments. Higher pay rates in the US were offset by allowances and welfare services in the UK, so that labour costs were much the same in the two countries.

Skill levels were not substantially different but supervisors had more responsibility and authority in the US. Both operatives and supervisors were aware of the need to produce good results; if performance was below the standards expected, jobs could be lost. Good workers were retained by supervisors for the more difficult jobs and often recruited from some distance because of their known abilities.

Based on report of the NEDO study,
Financial Times, *© F.T., 7 March 1990,*
reproduced with kind permission.

B.

The NEDO report *Faster Buildings for Industry* (HMSO, 1983) was based on 56 detailed case studies of building projects for industrial clients. The following extracts are taken from the concluding sections of the report.

On all 'fast' projects the design documentation – drawings and specifications – gave the contractor a well-defined basis to prepare and provide for the resource needs of the construction project... Programmes were based on critical path analyses and designed to open up and use work stations at the earliest opportunity. On smaller projects contractors used flexible multi-disciplinary teams. Progress was monitored closely and the contractor was ready to reschedule the work, using additional resources if necessary, in order to catch up with minor delays ...

The buildings were generally simple, designed with particular attention to ease and speed of construction; alternatively some concessions to buildability had been incorporated during preparation: these concerned mainly interference between trades, subdivision of the work into self-contained packages, concentration of services into particular areas, use of pre-finished components and types of finish which would reduce the congestion of trades towards the end of the project.

Programmes proposed by contractors tended to be designed with minimal site supervision in mind. Unless pressed for it, contractors had no good reason to offer shorter site times and increase the management risks and costs of speeded-up working ...

Where contractors were asked for both competitive cost and time, the tender results did not suggest a direct link or trade-off between construction times and cost: often the same tenderer offered the lowest for each.

WORKSHOP

1 In country X the average house requires 1500 man-hours to build. In country Y it takes 1800 man-hours to build a house, therefore productivity in X is nearly 17% higher than in Y.

- ***a*** What assumptions are implied by the statement that productivity is 17% higher in X?
- ***b*** Suggest three reasons why labour productivity might be higher in X than in Y.
- ***c*** Does higher productivity in X mean that they are more efficient in their use of resources than Y?

2 Describe and explain the variations in British and American productivity illustrated in Fig. 15.1.

3*Analyse the main causes of low productivity in the construction industry.

4 Evaluate the claim that local authority direct labour organizations achieve lower productivity than private sector contractors.

DISCUSSION QUESTIONS

- ❑ Who do you consider has the greatest responsibility for productivity: operatives, unions, site supervisors, management, government, others (specify)? Justify your choice.
- ❑ Which of the following do you regard as the greatest impediment to improved productivity: inadequate skills, poor communications, insufficient planning, poor design and/or documentation, other (specify)? Explain your answer.

16 The workforce

PREVIEW

- Why do some sites have labour problems and others not?
- What are the reasons for labour-only subcontracting becoming so widespread in recent years?
- Is subcontracting an efficient way of meeting the industry's labour requirements?
- What are the problems in planning for future labour requirements?
- How can the industry make more effective use of its workforce?

Comparative studies show UK sites employ more operatives than their foreign counterparts and take longer to finish contracts. There is less capital per worker and the proportion of skilled and supervisory staff to unskilled labour is also low. The industry relied heavily on casual labour in the past and relies increasingly on subcontracted labour today, which has engendered a hire-and-fire mentality that is at odds with planning the most effective use of human resources.

Construction is a labour-intensive industry and is likely to remain so. This makes it important to attract and retain a high quality workforce and to use it properly. People choosing a career look for good training, good prospects and good remuneration. Working conditions on site cannot be controlled to the same extent as in most manufacturing industries, but a regard for workers' safety and comfort is just as important. Firms with a reputation as a good employer reap the benefit in more commitment from their staff and improved productivity. Building good relations with subcontractors produces similar benefits.

WORKFORCE PLANNING

The concept of planning and implementing changes in the workforce as part of business strategy is unfamiliar to the building industry. Uncertain workloads and shifting locations militate against planning, but the costs of a constantly changing labour force can be considerable. They include the administrative costs of recruiting and interviewing. In addition there is the less obvious cost attached to lower productivity because workers have to start again with new colleagues and new management with each new job.

Good planning should cut the costs of obtaining labour and improve the use of existing labour. It includes establishing a proper ratio of labour to plant, monitoring the levels of training and skills available and anticipating labour requirements as far as possible, to avoid having lay-offs rapidly followed by recruitment. Where a firm employs its own labour directly these factors will influence decisions about planning workloads. With subcontract labour there is less concern about keeping labour employed although many firms regularly use the same subcontractors and try to keep some continuity.

Planning is about thinking ahead and trying to match resources to needs. Firstly, the workload must be assessed in terms of both the volume of work and the types of jobs to get some estimate of labour requirements. Then the labour resources currently available should be reviewed, their skills, ages, the rate of turnover, numbers of office and supervisory staff, etc. Whether additional recruitment is needed should then be considered, or further training, changes in the balance of skills, or even a reduction in numbers. Labour needs should be considered in conjunction with plans for investment in capital equipment.

Monitoring labour turnover is a useful aid to planning. The number of leavers per annum is calculated as a percentage of the average number of employees. Variations over time, or between sites, or even across companies, are an indicator of a firm's ability to retain workers. A high or rising turnover rate is a danger signal. It may be due to internal factors – pay, conditions, prospects, site supervision, etc. – or it may be the result of external factors such as alternative job opportunities, difficulties with transport or accommodation, etc. Finding out the reasons will help the firm to remedy the situation.

The extent to which planning is a formal task depends on the size of the firm. The number of very small firms with 2–3 employees more than doubled during the 1980s, as did self-employment. At the other end of the scale the number of firms employing over 1000 people remained more stable, but the total number of operatives employed by larger firms declined (see Table 16.1). These trends are partly due to the growth of subcontracting already noted, but they have serious implications for planning workforce numbers and for training. Small firms are more likely to react to market changes than to anticipate them. If larger firms are devolving their responsibilities for labour onto subcontractors, the industry as a whole must become more involved in forecasting future needs and implementing recruitment and training programmes.

Table 16.1 Employment of operatives by size of firm

October each year:	**1981**	**1983**	**1985**	**1987**	**1989**	**1991**
Size of firm: (employees)		**No. employed (thousands)**				
2–3	23.7	43.1	36.7	42.4	54.3	46.2
4–7	57.9	74.3	61.2	63.8	61.0	48.7
8–13	65.2	50.9	49.6	48.9	39.8	32.5
14–24	73.2	66.6	60.8	59.1	54.1	42.6
25–34	40.4	37.7	33.2	32.9	31.5	24.4
35–59	59.5	54.7	49.7	50.8	47.4	34.5
60–79	28.0	28.7	22.2	23.0	22.5	18.1
80–114	29.5	29.6	28.5	26.1	27.7	23.4
115–299	76.8	71.2	63.1	62.8	63.3	47.2
300–599	47.2	43.3	39.8	40.2	40.3	34.4
600–1199	45.6	42.6	38.5	41.7	35.0	27.9
1200 and over	79.6	69.1	65.5	57.2	73.6	57.6
All firms	**626.6**	**611.8**	**549.2**	**548.7**	**550.5**	**437.5**

Source: *Housing & Construction Statistics*, HMSO, 1993. Reproduced with the permission of the Controller of Her Majesty's Stationery Office.

MOTIVATION

Surveys have shown that management and operatives often have different perceptions of what people want from their jobs. If a high turnover of labour is due to operatives leaving a site because of worries over safety, bigger bonus payments are not the best solution. What motivates people may be quite complex and not always clear, even to the individuals concerned.

Sociological research suggests a hierarchy of needs, of which the most fundamental are material needs. Once we have satisfied these and have enough money to ensure a comfortable standard of living, other needs may take precedence. We want a satisfying social life and we want a sense of self-fulfilment, both of which we look for in our work as well as outside it. Payment is a form of recognition and helps satisfy these needs, but a word of praise can also boost morale (and productivity!).

Traditionally the construction industry has relied on incentive payments to provide the extra carrot. There is a bewildering assortment of different schemes, from all-out piece-rates, where payment is wholly tied to output, to schemes where the bulk of earnings is a basic wage, with some bonus related to performance on top. Often bonuses come to be regarded as a right and are even incorporated into the basic pay structure, which loses their point as incentives. Other schemes involve such elaborate targets and calculations that they are likely to cause disputes and do more harm than good. To be effective a scheme should be simple to understand and to calculate, related to work that is completed within a normal pay period, usually a week, contribute a sizeable proportion of the pay packet and be paid promptly.

Incentive payments are not suited to all circumstances. Day rates may be more appropriate where quality of work is more important than speed or if it is difficult to measure output, as with much repair and maintenance work. The targets set must be realistic, based on the capacity of an average worker exerting reasonable effort. If they are beyond the reach of most workers the incentive is lost. Evidence suggests that where bonuses are used their effectiveness in speeding the job is overestimated by both management and workers. Training, morale and the organization of the work are probably more significant than specific incentives.

SUBCONTRACT LABOUR

Labour-only subcontracting means self-employed operatives, either individually or in gangs, hire out their labour without becoming employees. It has become widespread especially in the south where many firms now retain only a core team of skilled workers on their permanent payroll. Although labour-only subcontractors can be paid on an hourly basis, in the majority of cases a price is negotiated for the job, which provides a similar incentive to piece rates to complete the job without delay.

Subcontracting is a rational response to many of the problems encountered in construction work. It makes the most of a scarce resource, skilled labour, by allowing self-employed tradespeople to work for different companies. It cuts down travel and transport costs because workers can move from job to job within their own locality instead of having to travel long distances to the employer's next contract. It encourages specialization and the practice of specific skills. From the main contractor's point of view there is the further benefit that some of the financial risk is transferred to the subcontractor and when workloads are low there is no long-term commitment involved, no redundancies to pay.

The disadvantages of subcontracting centre on the reduction in control and flexibility. The planning of a project, already a complex task, becomes that much more difficult. It also contributes to worries about poor safety standards and working conditions. With operatives in a hurry to finish the job and get on to the next, risks are sometimes taken which affect safety and quality. Quality in the longer term may suffer. Many labour-only subcontractors are small groups of self-employed workers who do not have the facilities, or time, to train the next generation in the skills of their trade.

TRAINING AND SKILLS

Training today is in a state of transition. While traditional craft skills remain important new products and new technologies call for a flexible, multi-skilled labour force. The old system of apprenticeship has declined. The growth of subcontracting has led to larger firms taking less responsibility for training. Widespread concern over the loss of traditional training opportunities together with recognition that a new approach was needed, led to the setting up of the Construction Industry Training Board. This is funded by compulsory contributions from all employers who can then reclaim part of the fees when they sponsor employees on training courses.

Varying three- or four-year schemes have been tried, which combined college training, in block or day release, with work experience and aimed at City and Guilds craft certificates. Now more flexible vocational training is being developed. Since 1983 the introduction of Youth Training Schemes has attempted to provide initial training and work experience for large numbers of school leavers. The programme has been widely criticized, not least for lacking in proper training content. Currently the trend is to try to involve industry more closely and to match training to employers' requirements through a graded system of national vocational qualifications (NVQs).

Comparisons with France or Germany, taking into account the relative size of the industries in each country, show the UK as having substantially fewer trainees gaining craft qualifications. Comparing quality of training is more difficult but two observations are worth noting. Firstly, the continental approach favours a broader training, which gives trainees more flexibility in their trade and a better appreciation of related tasks. Secondly, the on-the-job element of training is more closely supervised in those countries, where employers are required to provide trainees with a progression of tasks to develop their skills and to keep records of the work they have undertaken. This coordination of college and work experience is not sufficiently emphasized in the UK. Apprenticeship too often meant acting as a general dogsbody and college students had little contact with their tutors during industrial placements.

Training and education are not awarded high priority in Britain. Construction suffers additional handicaps. Its poor image does not attract many high calibre recruits. Too many entrants come to the industry because they cannot find anything better. Uncomfortable and hazardous working conditions, insecure employment and poor prospects are cited against the industry. Entrants who have made a positive choice are attracted by the outdoor environment, the variety and independence offered.

From the firm's point of view, training is expensive. Larger firms, working on contracts with tight schedules, may find trainees little more than a nuisance. The skilled workers, especially if they are on bonus, are reluctant to waste time supervising a learner. Consequently a disproportionate amount of training takes place in medium sized firms. Many of these are engaged in repair and maintenance work which gives a good variety of experience, but the large number of very small firms in the industry offer limited scope for training. The cost of training, even with a subsidy from the industry training levy, is something many smaller firms feel they cannot afford. In the UK relatively high wages are paid to trainees compared to fully qualified workers which again discourages employers from taking them on.

Improvements are being made. Certification of skills, introduced on a limited basis in 1983, aims to establish recognized levels of achievement through practical tests. An early example of this approach was a graded series of skill certificates for scaffolders; proven competence at each level gives operatives a certificate which recognizes their fitness to perform particular tasks. Basic skills and knowledge are built up in progression as on-the-job experience is incorporated into the training scheme. At the same time growing numbers of college students and academic courses reflect the increasing complexity of construction work.

Despite the difficulties highlighted here, considerable progress has been made. In the 1960s and 1970s it seemed inevitable that any major contract would over-run on time and budget. On one of the most notorious sites, the Isle of Grain power station, it was calculated that workers were being paid for an 8-hour day, out of which 6½ hours of productive labour could have been expected but an average of just 2½ hours a day were actually achieved. Today clients have a reasonable expectation of getting the work done as planned and some contractors' hoardings proudly announce completion dates ahead of schedule.

CASE STUDIES

A.

Two surveys undertaken from Loughborough University, one in the late 1970s, the other in the early 1980s, reveal a difference in perception between operatives and management concerning incentives to effective working on site. The rankings of the two surveys are given below; not all the items which appeared in the first survey were repeated in the second. (Top priority = 1.)

	Operative rating	Management rating
Earnings	3	1
Short travel time	7	—
Safety/working conditions	1	7
Welfare conditions	2	6
Job security	18	4
Friendliness of site	4=	10
Working as a team	12	9
Well organized site	4=	2
Good relations with management	14	3
Fringe benefits	15=	8
Recognition from management/workmates	10	—
Successful company	18	—
Modern company	15=	—
Challenging work	17	—
Job freedom	9	11
Plenty of time for personal life	6	5
Promotion prospects	21	12=
Training opportunities	20	12=
Ability to use/develop skills	8	—

Reproduced from Building, Technology and Management, *May 1984, with permission from the Chartered Institute of Building.*

B.

Montrose Construction Services is one of many recruitment agencies that specialize in the construction industry. In a four-week period it usually fills temporary vacancies for 450 trades and 250 technical people. Their contracts range from a few days to a year.

Enda Nolan is one of Montrose's temps, He says 'The thing I like most about temporary work is the flexibility. If I want to go to France for a month I can, and I know that I will have work when I get back. If I was asked what the three things were which attracted me to this type of life I would say flexibility, money and the ability to gain a wide range of experience.'

Peter Marjoram, another of Montrose's temps, started out as a farm manager. He then became apprenticed to a small builder 'who taught me carpentry and joinery. It wasn't a formal apprenticeship, but I learnt a lot.' Employment with various construction companies followed before he decided to freelance. 'I do it this way for the money. The building business is about money and money is made and lost on site. The site agent is responsible for running the job. Also I can make my own decisions about pensions and provisions for my retirement. The best thing about temporary work is that I don't have to get involved in petty office politics, the constant jockeying for position and the backbiting.'

Reproduced with permission from Building Today, *23 June 1988.*

C.

The effects of inadequate standards of training in building trades have become a matter of widespread concern. Both inside and outside the industry tales of incompetence are rife. The number of defects that result from poor workmanship is worrying to clients and damages the reputation of the industry. They range from small-scale repairs badly done by local builders to serious failures in many modern high-rise blocks, which have led in some cases to collapse or demolition of the properties. Personal evidence of incompetence can be gathered from almost every site, often laughable, sometimes frightening, occasionally bizarre.

One such is the case of the householder who arrived home to find his kitchen door would not open. The reason, he found, was that while he was out his kitchen had mysteriously filled up with concrete. There were no builders in his flat but there was an office block under construction next door to him. Further investigation revealed that workmen on the site had treated the wall of his flat as 'permanent shuttering' and poured concrete for the new offices against his property. No-one realized that the cavity brick wall was collapsing under the sideways thrust of the weight of concrete against it ... The contractor thought the engineer had given authority for the work, the engineer understood ... etc.

Based on a paper by S.J. Prais and H. Steadman, Productivity, Education and Training, *NIESR, 1990.*

WORKSHOP

1 Comment on the survey results shown in case study **A**.
Would you expect a similar survey undertaken today to produce similar results? Explain any likely changes.

2 Discuss the pros and cons of widespread use of agency labour on sites.

3 Discuss the benefits of bonus incentive payments and indicate the main problems encountered in setting up an effective scheme.

4 Frequent dissatisfaction has been expressed with the state of training in the industry. What do you consider to be the main impediments to better training?

5 Consider the case of the concrete kitchen (case study **C**). What do you think was the cause of this calamity? What remedy would you suggest to prevent a recurrence?

DISCUSSION QUESTION

❑ Compare the most efficient contract you have worked on with the least efficient you have experienced. Outline the main differences between them. Does this lead to any conclusions which would assist in the effective use of manpower?

17 Capital equipment

PREVIEW

- What sort of resources make up a firm's 'capital'?
- What determines the choice between labour intensive methods or capital intensive methods of building?
- Why is there less mechanization in building than in manufacturing?
- Why will one firm choose to buy plant and another firm choose to hire?

WHAT IS 'CAPITAL'?

This chapter is headed 'capital equipment' rather than 'capital' to emphasize the fact that capital is a factor of production: physical assets which are used in the production process. Fixed capital refers to assets which can be used repeatedly, such as buildings, plant and vehicles, whereas working capital are the assets which can only be used once in the course of production, such as stocks of components and materials, bricks, concrete, chimney linings, bathroom fittings, etc. Cash in the bank is also capital in liquid form, meaning it represents resources which are available but not yet committed to a specific use.

We saw in Chapter 6 that 'capital' has more than one meaning. In financial terms a company's capitalization, or capital value, means the market value of the business. This is not the same thing as the total value of its physical assets or capital equipment. The value of the company is its value as a going concern, a business capable of yielding an income. Roughly speaking it is the share price multiplied by the number of shares issued, or what it would cost to buy the company. A profitable firm with a good order book will be valued more highly than a firm whose current contract is running at a loss and with no new jobs in prospect. Share prices or capitalization will be quite different for these two firms, but they may be very similar in respect of their physical asset values, their holdings of property, plant and stocks.

Here we are concerned with capital as a resource, assets contributing to production. The distinction between fixed (illiquid) and more liquid assets is not a rigid one. Working capital, in the sense of liquid reserves or money, is 'tied up' when it is used to purchase fixed assets, such as concrete mixers, but a firm may opt instead to hire mixers as required, substituting the use of working capital for fixed investment. Decisions about the use of capital involve choosing the right capital goods (plant, equipment) for the business and deciding the right way to fund them.

CAPITAL INVESTMENT IN THE CONSTRUCTION INDUSTRY

The construction industry uses less capital than manufacturing. On average a construction employee has only half as much plant and equipment to assist him as the average factory worker. The ratio of capital : labour depends on a number of factors. These include the extent to which firms are forced, by competition, to adopt the most efficient methods available, the productivity and price of capital compared with that of labour, and the organization of the industry.

The organization of the construction industry, with its large number of small firms and those who are self-employed, reduces the overall level of capital investment. Small firms generally lack the financial resources or the technical knowledge to buy the latest equipment. Except for specialist sub-contractors, varied workloads make it uneconomic to buy more specialized pieces of equipment. Even if the capital is non-specific (i.e. capable of a wide range of applications) construction firms seem slow to invest. Computer technology has been gaining ground steadily but the industry spends only 0.25% of turnover on it, far less than other industries.

Competition is variable, as we have already seen, owing to the fragmented nature of the market. Imperfect knowledge and geographical limitations insulate many smaller firms from competitive pressures to some extent. Larger contracts account for approximately half the market in terms of value. The effectiveness of tenders in bringing competition to bear has already been discussed. Bids can vary quite widely, but informal contacts keep most firms in touch with the prices being offered by their competitors.

The scope for mechanizing construction work is limited in various ways. Site work involves a series of different, consecutive tasks rather than constant repetition of the same task. Mechanization is easiest to achieve where a task is repetitive and can be closely specified. Tasks that require skills, such as plumbing or bricklaying, are less easy to mechanize than routine work. A large proportion of building work is repair and maintenance, where access may be tricky and each job is different. Labour has the fundamental advantages of adaptability and judgement. For these reasons traditional building has remained relatively labour intensive.

SUBSTITUTION OF LABOUR BY CAPITAL

Theoretically the ideal balance between labour and capital can be expressed by the equation:

$$\frac{\text{Marginal physical product (MPP) of capital}}{\text{Price of capital}} = \frac{\text{MPP of labour}}{\text{Price of labour}}$$

If capital becomes more productive (through advances in technology) or cheaper (through price reductions or lower finance charges) it will tend to replace labour. A rise in labour costs, or reduced labour productivity, would have the same result. Historically trends in labour markets and advances in technology have tended to encourage the use of plant in place of labour, although the rate of substitution has been uneven. The 'navvy', an unskilled labourer who worked with pick and shovel, has all but vanished in the last thirty years. A combination of rising wage costs and machinery that can shift tons of earth in a matter of hours has rendered the navvy uneconomic, his labour substituted by capital. In practice investment in machinery also depends on the extent to which it provides a convenient solution to the job. The large excavators that have displaced the navvies from road construction are not practical or economic on smaller sites and it is only in recent years that more manoeuvrable machines, mini-excavators, skid-steer loaders, etc. have taken over in the more confined spaces.

Rising labour costs and shortages of labour promoted new building methods in the 1960s (see the section below on industrialized building methods). Population changes suggest we may be about to experience labour shortages again. Even with present high levels of unemployment skill shortages can hamper the industry. Rapid growth in the field of computer technology and the development of intelligent machines capable of monitoring and adjusting their own performance may transform the labour market. Automation and robotics have had little impact on construction to date, but research is progressing. There are already robots capable of simple bricklaying or finishing concrete floors. These are single-task machines and as yet their performance is no better and the cost higher than doing the job by hand, but rising labour rates could tip the balance. A machine which can crawl across vertical surfaces, employing sensors to monitor cracks and damage to walls, sounds like science fiction but is reality. The potential productivity gains here are much greater because of the costs of scaffolding and risks to human life of working on high-rise buildings.

Where capital has replaced labour most readily to date is in operations where sheer power is the main requirement. Excavations, lifting and transport are all undertaken more efficiently with the use of bulldozers, scrapers, cranes and motorized vehicles. This sort of plant is non-specific, it works in a variety of situations. Adaptability remains important in site work. Where capital is used on a large scale, its use is associated with radical changes in design and methods of erection, such as the high-rise buildings of the 1960s, which utilized tower cranes and hydraulic pumps capable of delivering concrete to a height of 300 m.

Plant is expensive; if it spends most of its life in the yard it represents idle resources. It must be used if it is to earn a profit. Low utilization rates occur because of unforeseen delays due to bad weather or unexpected soil conditions, or through poor planning and site organization. Utilization rates vary widely, from as little as 25% of the available time for items such as water pumps to 70% for excavators. Site preparation is more highly mechanized than many other activities partly because it takes place in a single uninterrupted phase, allowing machinery to be moved in, used and moved on to another location. As more fixed capital is employed it becomes more critical to ensure it is fully used. At the same time higher utilization rates increase the productivity of capital and encourage further investment.

The trend towards subcontracting specialist tasks is one rational response to the problem of low utilization rates. Specialist firms can justify the purchase of specialist equipment because they will be able to use it to the full. An alternative approach to getting more intensive use of plant is to hire it as it is needed. The choice of plant is also increased in this way, since hire companies can offer a wide range.

PLANT HIRE

Plant hire, a post-war phenomenon, has expanded considerably in the last twenty years. Smaller firms gain the use of equipment they cannot afford to buy. Larger companies have the chance to off-hire plant that would otherwise be standing idle. Less obvious benefits arise from the opportunities for introducing and appraising new plant. Hire companies have an incentive to inform potential users of the advantages of their equipment, and they receive feedback on its merits and drawbacks from a wide range of users. New items like the Komatsu mini-excavators which were introduced in the early 1980s have penetrated the market more rapidly as a result of their capabilities being experienced on hire. Larger, more expensive and specialized items, like hydraulic concrete pumps which are only required for short spells, may only be economic if hired.

The choice between hire or purchase is essentially a matter of opportunity costs, comparing alternative ways of using scarce resources. For the occasional, one-off job hiring has many advantages. Initial outlay is less. Defective plant will be replaced without extra expense. The firm avoids the overheads associated with owning plant – depreciation, maintenance, off-site storage facilities, administration and interest on finance. Costing the plant element in the estimates is simplified. Other elements to consider are the equipment's potential lifespan and the risk of obsolescence.

Purchase is indicated if utilization is likely to be consistently high. Although the initial cost is greater, costs per job will be less. Owned plant is constantly available, operatives are familiar with its use and the firm has control over maintenance. Regular maintenance is important to gain maximum benefit from equipment. Breakdowns can be very expensive if other work on site is delayed in consequence. Of course maintenance adds to the costs of owning plant; the profit maximizing principle of MC = MR is once again the guiding principle. In this case the marginal cost of extra maintenance must be balanced against the savings from fewer breakdowns. Records of maintenance work provide data to establish the balance.

Off-hiring plant when it is not in use is an additional source of revenue but not an end in itself. There is always a risk that off-hiring restricts the company's own use of the equipment. Profitable use of plant relates primarily to its contribution to efficiency in undertaking contracts. Another danger unattached to ownership is the temptation to use plant because it is there, even though it is not the best equipment for the job in hand. The versatile bulldozer is still used for many jobs where a more specialized excavator or grader would be better.

Where plant is required for longer periods, but a firm does not want the capital expense of buying equipment, contract rental or leasing can be considered. Contract rental is essentially a fixed-term hiring. It gives the hirer guaranteed use of plant without the responsibilities of ownership. Leasing is an arrangement based on the separation of ownership and control. The user gains sole and unrestricted use of plant without gaining title to it. Use normally includes responsibility for servicing and other overheads in this context, as the leasing company's role is that of providing and financing the plant only.

In many cases the deciding factor is financial. Purchase of plant, even when justified in the long run, may require an initial outlay which is difficult to cover. In these circumstances the high marginal cost of hiring may be preferred to purchase, even though the latter has a lower marginal cost in use which makes it cheaper in the long run. Smaller firms are particularly vulnerable to budget constraints.

INDUSTRIALIZED BUILDING METHODS

Whilst plant hire aids productivity through a more rational use of capital it has not substantially altered the capital : labour ratio in traditional building. Mass production methods, where work is transferred from site to factory, implies considerable extra investment. Economic and technical factors contributed to this development.

To make mass production worthwhile there must clearly be a mass market. Widespread destruction of housing during the Second World War left a massive shortage of accommodation. Prefabricated dwellings were the quickest solution. Over a hundred systems were approved. Many used concrete, either as pre-cast sections or poured *in situ*, but steel and timber-frame were also adopted. Generally these temporary houses were not popular, they lacked aesthetic appeal and they were very expensive. Their main virtue was their speed of erection, but many of them also proved surprisingly resilient and survived well beyond their intended lifespan. After 1947 grants were withdrawn and prefabs gave way to a return to traditional styles.

By the 1960s housing was again in crisis. Much of the nation's housing stock was old and dilapidated, yet incomes and expectations were higher than ever before. Construction output was not keeping pace with the community's needs, nor was the industry likely to do so in an era of full employment and labour shortages. Government policy encouraged local authorities to speed up their housing programmes by using industrialized methods. New systems proliferated. Unfortunately their very numbers reduced the potential advantages of mass production.

Two main categories can be distinguished. Closed systems, usually designed by a contractor, used components designed as part of that specific system. Open systems relied on modular designs and standardized components to create an off-the-shelf range of buildings designed to the client's specifications. Some systems, such as CLASP (Consortium of Local Authorities Schools Programme), were client based and provided designers with a catalogue from which to work.

The potential benefits were considerable. Big savings in site time meant work suffered fewer delays from bad weather. Non-productive tasks, such as scaffolding, were cut back and there was less wastage of materials. Factory production of components meant easier quality control. The use of a tested and proven system could release design time from work on the basic elements of the structure to concentrate on customizing the building to meet the needs of the individual client. In all, the new systems promised higher productivity, increased speed and better value for money. From Easterhouse in Glasgow to Redbridge in Southampton the tower blocks reared up, a monument to planners' aspirations.

Today those same blocks are more likely to be seen as monuments to folly. Of course not all systems buildings were high-rise and much of the criticism of the tower blocks revolved around the social isolation they imposed on young families, but there were other faults which were not restricted to the high-rise buildings. Twenty years on many of the houses built by the new techniques have been demolished or have required substantial repair work.

The most common complaints relate to water penetration; the most serious involve major structural failure. In some cases design was at fault. Detailing that looked sound on the drawing board gave inadequate protection in practice against wind and rain. Other failures arose from bad workmanship. Construction workers were not properly trained to appreciate the need for exactness which the new methods required. Ignorance and poor supervision resulted in fastenings being improperly fixed, or even left out, concrete being improperly mixed with consequent loss of strength, and a host of other faults.

In many cases technical problems did not appear until later but by the 1970s opinion had already turned against these buildings. The promise of good, fast, inexpensive housing had not been fulfilled. Changing economic circumstances were part of the problem. Costs were expected to fall as the industry gained experience with the new methods and as the high initial investment was spread over larger outputs. The early expectations were not met, savings proved disappointingly slight.

Too many competing systems meant too small a market for each. Labour shortages gave way to unemployment and consequently labour costs did not rise as fast as predicted. Costs of materials and transport, however, rose more rapidly after the oil crisis of 1973/4. Thus the substitution of capital for labour gave less cost advantage to the new methods than expected. Cutbacks in government spending filtered through to local authorities and the volume of large contracts dwindled. The emphasis in housing policy switched to encouraging owner occupation. Owner occupiers and their financiers, the building societies, favoured low-rise, traditional housing. The experiment petered out.

The mistakes that were made do not invalidate the ideas and technologies involved. Greater standardization of components and modular design can reduce costs even with traditional materials. Timber-frame housing combines many of the features of industrialized building with a more traditional appearance and a style of dwelling which housebuyers prefer. Systems such as CLASP, for school buildings, proved popular and successful. Fully integrated systems still have much to offer, especially for buildings in the commercial and industrial sectors where time savings are a significant factor. Widespread use is made of prefabricated elements, such as toilet modules. It is easy to envisage circumstances in the future which could once again push the industry towards further application of mass production methods.

CASE STUDIES

A.

'Machines have become more versatile and labour saving,' confirms Matbro's Chris Morley. 'Plant has developed into smaller equipment to work in urban areas.' Jobs in confined spaces, traditionally done by hand, can be mechanized now with micro-loaders and excavators that can be craned into coffer-dams, basements or work within tunnels.

There has been the same lift-off in materials handling on site: first with the introduction of the rough terrain forklift and, more recently, with the telescopic handler. 'Basically, it is a labour-saving device looked at with some scepticism by brickies,' says Morley. 'You would previously have had to hire a crane or hoist to load out. With its front-end attachments, it is a more flexible piece of equipment, is doing a variety of jobs, and must be eating into sales of other equipment.'

New materials are being incorporated into wear parts which adds up to fewer grease points and reduced maintenance. 'There is longer life in components and designers are examining all those places where steel grinds against steel,' says Morley, 'to see if these new materials can change the timescale of repair.'

Reproduced with permission from Building, *1 May 1987.*

B.

How do the hirers themselves make a profit out of plant? Vibroplant director Harry Stiano acknowledges that the initial purchase price is very important. 'Equally important is what you buy and who you buy it from,' he argues. 'If I had to buy on price alone, I would have a very different fleet to what I've got today. I buy on a combination of the right machine for the job and the right specification. The commercial aspects of the deal are important but our highest priority is who we buy it from, their past and current history. There's no point in going out and doing a deal with someone who looks so shaky he'll be bust next year.'

Once in the fleet, whether it's one week old, one year old, or five years old, if it comes within the maintenance plan – and there are not many items that don't – the equipment is checked over on a four-week cycle. 'We spend an awful lot of money on preventative maintenance,' says Stiano, 'an awful lot of money on running repairs, and a major amount on major repairs.' Speed and ease of maintenance are key assets. 'It's no use buying a machine that takes a day and a half to fix the clutch,' he explains. 'Having said that, and got everything right in the buying equation, if it's unacceptable to the client, then I've not done a good deal.'

Disposal is a very important part of the total ownership equation. The secondhand plant market is overstocked but Vibroplant hopes to make a profit by either writing down correctly or making sure the item is in reasonable shape. 'We don't go out to sell it ourselves,' says Stiano, 'we haven't a selling organization. We invite dealers into our yard and they give us a price.'

Reproduced with permission from Building, *1 May 1987.*

C.

In 1965 the Ministry of Housing and Local government issued Circular No. 76/65, *Industrialized Building* (HMSO, 1965) 'to launch a concentrated drive to increase and improve the use of industrialized methods in housebuilding for the public sector.' The benefits which were expected included:

Numbers – 'this is the only way to build the number of houses we need.'
Speed – 'most industrialized techniques show worthwhile savings.'
Design – 'carefully prepared standard designs will release scarce professional time to concentrate on raising the quality of layouts.'
Quality – 'industrialized methods facilitate quality control.'
Price – 'efficient organization of supply and demand can bring down promotion costs.'

In 1980 the Department of the Environment brought out Occasional Paper No. 4/80, *An Investigation into Difficult to Let Dwellings*, HMSO, 1980. It was a report critical of a number of developments 'which are inhuman in scale, uniform and repetitive in appearance and inadequately provided with social and community facilities. An industrialized building system had been employed on 16 of the sample estates and many of these had massive concrete facades of overwhelming severty which sophisticated and generous landscaping (if provided in the first place and if it survived) could do little to mitigate.

'In addition to their often unattractive appearance, technical problems, such as water penetration, were not uncommon with the system built schemes we looked at.'

D

'It's just like Lego,' said Mr Perrin, 'because just as you can use Lego bricks to create things that look very different, so you can use the Tartan system to create very different buildings by altering the size of beams between the fastening nodes and varying the style of cladding.'

This is a system closer to engineering than to traditional construction techniques. It is based on a pressed steel frame produced to tolerances of + or - 1.5mm. 'This accuracy is essential to the Tartan design, because the frames are drilled with holes for the fastening nodes and the passage of services, pipes and cable, and once the basic frame is up everything else depends on it, so there can be no scope for bodging,' said Mr Perrin. The frame is held together by fastenings which allow universal connections so that standard modules can be combined into buildings of varying sizes and shapes. Cladding takes the form of enamelled metal panels, which hook onto the frame direct from the delivery lorry. Fastening is completed from inside the building.

Mr Perrin was motivated to develop the system by the frustration he felt when confronted with the waste of time and materials, and the inability to repeat and refine concepts in the conventional one-off contract. He also sees an advantage to clients, who can see what they're getting before they commit themselves.

Based on an article in Financial Times, © F.T., *4 February 1985, reproduced with kind permission.*

WORKSHOP

1 Explain the following terms and illustrate with figures based on your own company's annual report:
   a market capitalization
   b capital investment
   c fixed assets
   d liquid reserves
   e asset value.

2 Discuss how investment decisions relating to plant and equipment may affect the firm's labour requirements.

3 Plant-hire companies and construction companies are purchasers of construction plant. Compare and contrast the considerations which are relevant to their purchasing decisions.

4*What factors should be considered when deciding whether to purchase or hire a piece of equipment?

5 Find a local example of industrialized building methods. Describe the main construction methods used and assess how far the performance of the building could be described as satisfactory. What were the main causes of any problems experienced with the building?

6*Discuss the economic trends which could favour increased use of industrialized methods in building. Are there any other factors which might support these pressures?

DISCUSSION QUESTIONS

- ❑ What do you see as the most important new developments in construction equipment over the past ten years?
- ❑ How do you think these will affect the use of resources in the industry?

18 Land and site values

PREVIEW

- Why is a building plot in London worth more than a similar plot in Luton?
- Why are rents in the City of London square mile higher than rents in the suburbs?
- Do planning controls help or hinder market forces in making the most effective use of land?
- Why are some buildings demolished long before they become unsafe?
- Should historic buildings be preserved at all costs?

DETERMINING THE VALUE OF LAND

Building, like all forms of production, uses the natural resources classified by economists as 'land'. Above all building requires land in its primary everyday sense as space, a site on which to build. Chapter 4 examined how far the supply of land could be regarded as fixed. Land for building is certainly restricted. The supply, while not entirely fixed, is very inelastic and therefore prices will depend on demand for development.

How much a developer is willing to pay for a site will depend on how much they can get for the development when it is completed. Starting with the **gross development value**, an estimate of the potential selling price or value of the completed project, the calculation works backwards. Construction costs and developer's profit are deducted from the gross development value. The residual sum which remains is the amount available for land purchase. To offer more would reduce the scheme's potential profit; if it can be obtained for less the developer stands to make extra (abnormal) profit. An example is given at the end of this chapter.

Residual valuation, as above, is not just for calculating land values for new developments. Even a homebuyer who looks at a house priced at, say, £60 000 and judges that repairs and improvements costing £20 000 would raise the value to £90 000 and therefore decides to buy is using the principles of residual valuation. Whilst the method is simple, putting it into practice is not always easy. The homebuyer may find the cost of repairs greater than expected. Apparently minor defects may prove to be more serious. Renovation will in any case take some time to complete and by then market values may have changed. Prices of older houses fell by just over 5% in six months in 1992 (Nationwide Building Society index) which would have reduced the expected sale price to approximately £85 500 cutting the profit by nearly a half. With more complex developments, risks are proportionately greater.

If the buyer had not intended to sell the house, its market value after improvement would not be the main concern. How much an owner occupier spends depends mainly on the utility offered by the improvements and the available budget. If the buyer is acting as a developer and intending to make a profit, the resale price becomes crucial. Most development is undertaken for profit, relatively little is for occupation by the developer. Land is needed for development, or for other productive uses. It is rarely wanted as an end in itself. Demand for it is a derived demand, so the price of land depends on the demand for houses built on it, crops grown on it, etc.

Consider the movement of house prices around Aberdeen during the boom period of off-shore oil drilling. As the highly paid oil workers moved in, competing for places to live, house prices soared. High house prices stimulated building activity and land prices also rose. With exploration completed, the oil companies laid off their drillers and construction workers and moved their managers to new sites. The loss of jobs and incomes in the area caused a dramatic fall in house prices. Many who had purchased at the peak of the boom found themselves in the mid 1980s unemployed or having to take lower paid jobs, unable either to keep up the mortgage payments or to find a buyer for the house. Naturally land prices plummeted also. The qualities of the land had not changed – it was the demand for it that rose and then collapsed. Recession in the oil industry kept demand low and not until oil company profits recovered towards the end of the 1980s did the Aberdeen economy and its housing market pick up again.

The supply side of the property market is slow to alter, so demand sets the price of land over the short term. In a free market supply and demand not only set prices, but also determine what use is made of the site. Land is scarce and there are many potential uses for it: houses, shops, offices, factories, roads, recreation, farms and many other uses compete for sites. How much any one user is prepared to pay for a particular site will depend on how much profit they can generate – if housing will produce more profit than shops a housebuilder will be prepared to pay more than a retail developer. In a free market the land goes to the highest bidder, thus land prices and land use are determined simultaneously.

LAND VALUES AND LAND USE

Von Thunen, a German economist, in the first half of the nineteenth century showed how land use was determined by supply and demand. His model, based on the rural economy in which he lived, focused on the distance from the market at which various agricultural products would cease to be worth producing (Fig. 18.1). He made the assumptions that costs of cultivation and revenues from the sale of crops would not be affected by distance, so the only variable would be transport costs. These would depend on the nature of the crop and the distance from the market-place. Dairy farmers would suffer greater penalties from having to transport their goods over large distances because their goods were more perishable (transport consisted of horse and cart in Von Thunen's day) than wheat growers. Dairy farmers would thus be willing to bid more money for land closer to the centre than wheat growers. In this way land use and value were determined together.

Despite its over-simplifications, the Von Thunen model provided a good basis for developing theories to explain patterns of land use in what was

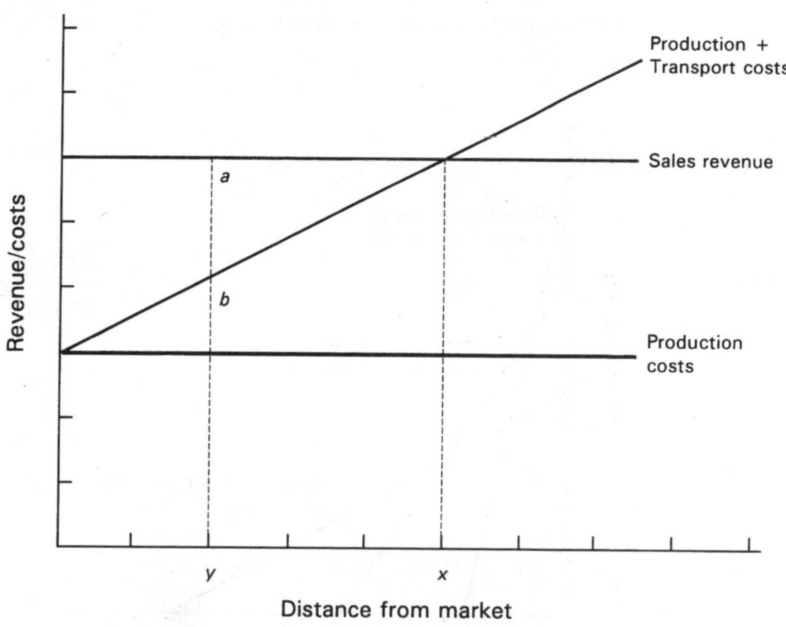

At a distance of *x* miles from the market production ceases to be profitable because of the rising transport costs.

Closer to the market, at a distance of *y*, the potential excess of revenue over total costs is *a*-*b*. This shows the maximum amount of rent that can be paid by the producer for land at this location.

Figure 18.1
Rent earning capacity after Von Thunen.

essentially a rural, market economy. More sophisticated models are needed for the land uses that have grown up around urban centres. In the 1960s Alonso introduced a broader notion of **accessibility** in place of simple transport costs and produced a rent bid model which is more flexible and more appropriate to urban land use (Fig. 18.2). Accessibility includes the advantages a location can offer in terms of easy access to resources, especially labour, as well as access to markets. Being close to resources helps to keep costs down, being readily accessible to customers helps to keep revenues up. As firms move outwards from the central business district (CBD), away from the most accessible locations, falling rent levels reflect a trade-off against profits. Being in an area of less accessibility raises the costs of running a business and reduces its earning potential; to compensate, the firm offers a lower rent for the more peripheral location.

As with the earlier, Von Thunen model sensitivity to location varies, so a distinct pattern of land use develops. Shops which need a high degree of visibility to attract as many customers as possible will lose trade rapidly as they move away from the centre. Their rent bid curves are consequently very steep. More specialist outlets have less need of passing trade and so less to gain from a central location. Light industrial use, warehouses, etc., have

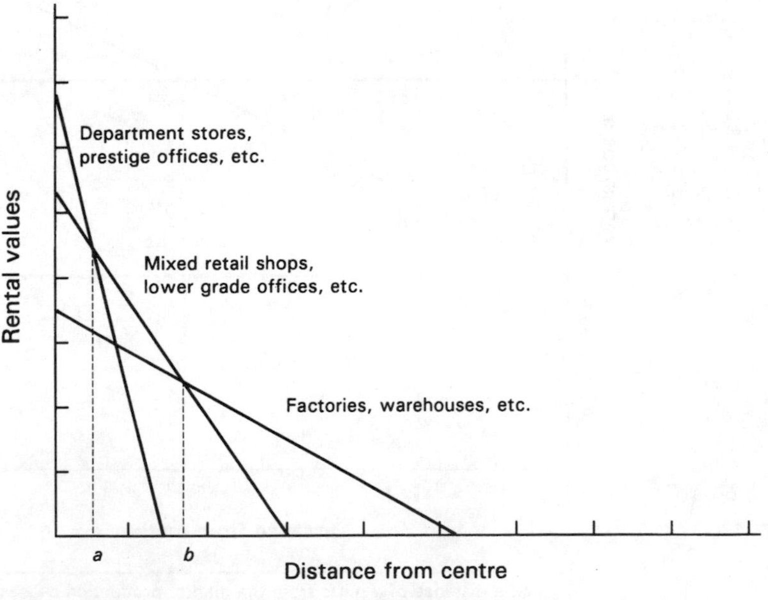

The central zone as far as *a* is dominated by businesses dependent on maximum accessibility. Between *a* and *b* businesses with lower rent earning capactity predominate. Beyond *b* are businesses for whom a central location is less critical still.

Figure 18.2
Rent bid curves.

much flatter rent bid curves which reflect less impact on their businesses from the location of their premises.

The city centre, the focus of transport systems, becomes its business centre, dominated by shops and offices – people-based businesses for whom accessibility is all-important (Fig. 18.3). Because the supply of sites in the centre is very limited and the demand for these is great, rents will be high and users whose profits are less affected by location tend to be squeezed out.

The high rents commanded by central locations means more intensive use is made of these sites. In order not to 'waste' the expensive land, offices are placed above the shops and developers build upwards to create as much usable space as can profitably be let on each site. On the edge of town, where land is cheaper, it can be used more extensively. Commercial developments are low-rise, houses built on more generous plots. The ratios of capital to land employed reflect their relative costs; by building upwards we 'substitute' capital for land, where land is in short supply and thus expensive.

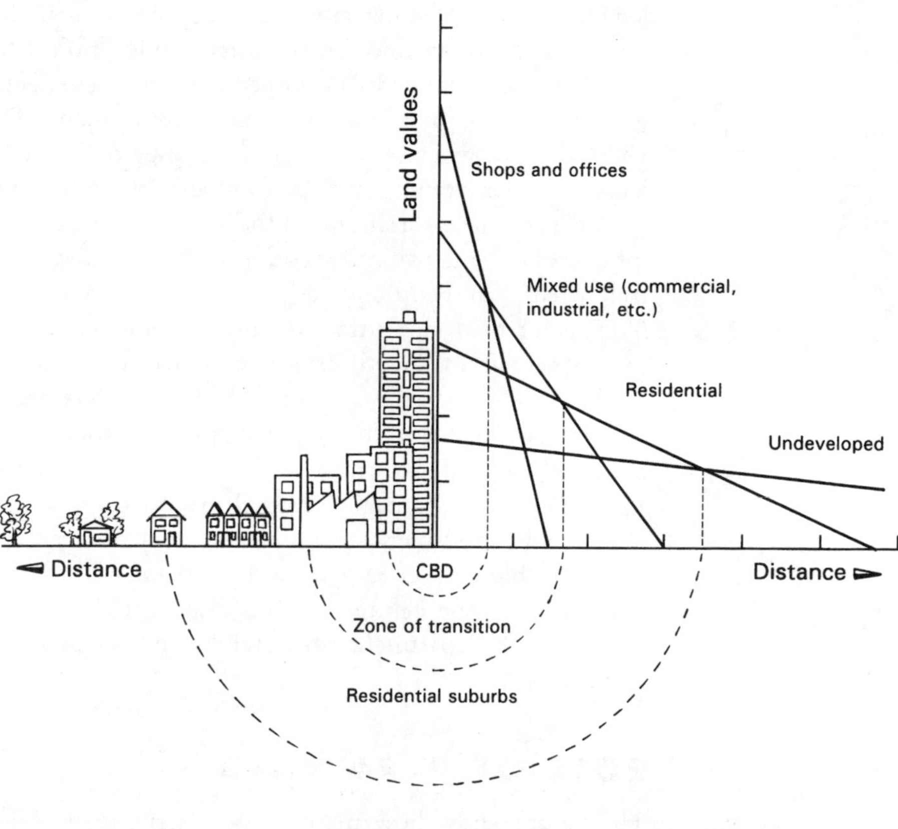

Figure 18.3
Nature and intensity of land use.

If accessibility is largely a function of distance the pattern of land use will develop in concentric rings around the centre. Moving outwards, each type of use gives way to another, as its ability to generate enough earnings to outbid the next best user diminishes with distance. The central business district gives way to a transitional area of mixed commercial and industrial activity before reaching the suburbs, the predominantly residential areas. Here householders balance the need for access to shops and jobs with the advantages of more spacious and peaceful surroundings.

The model is very schematic. In reality urban areas are dynamic, ever changing. Growth will push out the boundaries of each zone, but because buildings may continue in their existing uses for many years, different types of land use coexist side by side. Accessibility has been treated as broadly dependent on distance, but it must take account of local topography and historical patterns of development too. The concept of accessibility is itself changing with higher incomes, more car ownership and more leisure time. Many people find new out-of-town shopping areas more accessible than congested city centres. In larger urban areas the growth of the suburbs has led to problems of inner city decay, aggravated as traders move out to the new centres. As demand for the inner city locations falls, the area becomes less attractive to investors in property. Lower values lead to less willingness to spend money on maintenance and improvements. Dereliction leads to a vicious circle, whereby as the more prosperous residents move out, they leave the inner city impoverished and less able to halt the decay.

There are many variations on the broad pattern described and many local influences to be taken into account. In bigger cities, centres serving specialized needs can be found, as a particular pattern of use is established. Competing firms are attracted into locations close to each other, where they benefit from shared services or enhanced customer appeal. London's Hatton Garden became a centre for the jewellery trade where gem stone dealers, jewellery workshops and suppliers of tools to the trade congregated. Bond Street is known for fashions, Soho for entertainment, etc. These areas share the advantages of **agglomeration**, the grouping together of similar and complementary businesses which attract a larger number of customers. A big city serves a wider hinterland and so can offer more specialized services. The central business district itself becomes subdivided with clusterings of department stores, clubs and restaurants, lawyers and estate agencies, etc.

ROLE OF PLANNING

The models show how market forces serve to establish land use patterns through the price mechanism. Since the 1920s market forces have been modified by planning controls. These operate at two levels: through structure plans, outlining the broad pattern of development for an area, and individual consents for specific projects. Planning acts as a corrective where the market fails to achieve the best outcome. Markets work by people pursuing their own interests, often without regard to the effects on others. If the spillover effects are harmful the market fails to maximize welfare for the community as a whole.

Building changes the environment in ways which have widespread effects. The sale of derelict land to build a factory may well be welcomed by some as providing new jobs and clearing up an eyesore. Others may oppose the scheme because it means pollution, noise and extra traffic congestion. Whatever the costs or gains to local residents, they are external from the developer's viewpoint. Unless the residents have the power to influence the commercial success of the project their wishes will not affect its profit potential, which is the developer's main concern. Conflicts of interest which arise in these circumstances can be dealt with administratively, e.g. by planning controls which designate the type of development permitted in different areas. This separates incompatible uses such as heavy industry and housing. Planning can also be used more positively to encourage complementary uses, e.g. housing and recreational space. Alternatively the externalities can be brought into the market so that the price mechanism takes account of them. These issues are discussed in Chapter 20.

CHANGES OF USE AND REDEVELOPMENT

Within the limits imposed by planning restrictions, it can be assumed that developers will seek to maximize profits. A site will be put to the 'highest and best' use available, the one which yields the best return. Over the years changes in costs and revenues occur. Costs rise as more maintenance and repairs are needed. Eventually the building may become structurally unsound and may be abandoned or demolished. It has reached the end of its physical lifetime.

More often redevelopment takes place for economic reasons long before this point is reached. As costs rise, revenues fall. During the early years of its life the building commands good rents; its earning capacity may even rise if demand for accommodation outstrips supply. In time rising maintenance costs and dilapidations make it less attractive to tenants. The value of the building falls. At the same time the value of the site increases because of the shortage of land for development. Eventually a time is reached when the site value is greater than the existing value of the building, at which point it is ripe for redevelopment (Fig. 18.4).

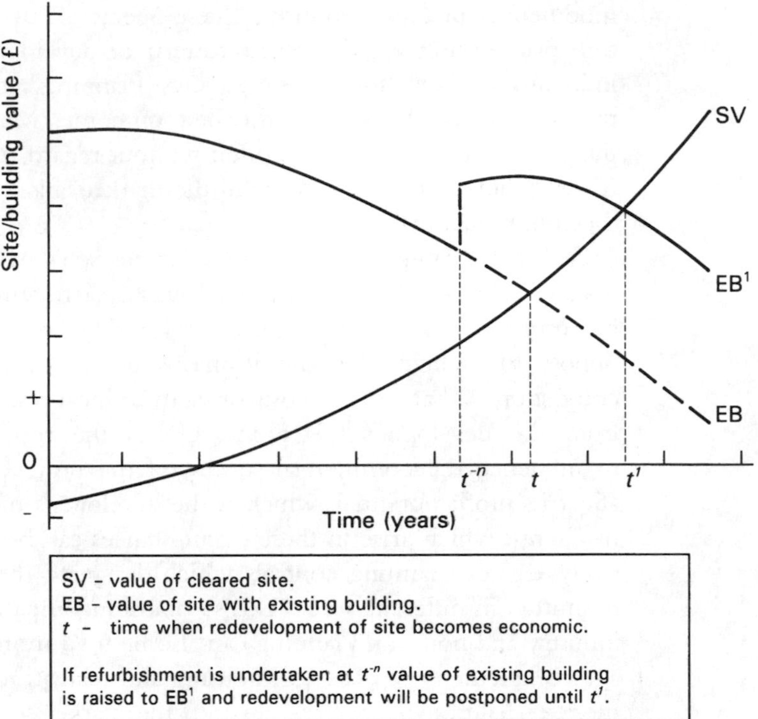

Figure 18.4 Economic lifetime of a building.

The building's economic life is not fixed by its physical or structural attributes, it is a matter of market forces. It is therefore uncertain and subject to change. Demand for existing uses may fall – textile mills fell vacant with the decline of the cotton industry and empty crofts bear witness to changed lifestyles in the highlands of Scotland. At the same time demand for alternative developments is increasing, whether for well-equipped office space or redundant farm buildings for conversion into holiday homes. When the potential value of the site, inclusive of demolition and rebuilding costs, overtakes its value in existing use, the present building has reached the end of its economic life.

Changes which lower the cost of redevelopment will shorten the economic life of a building. This includes falling interest rates, or government subsidies. Redevelopment is postponed if costs rise or if new building techniques and materials reduce the costs of renovation, making it cheaper to refurbish. As we become more concerned to conserve our industrial and architectural heritage, refurbishment is often preferred to total redevelopment.

The preservation of historic buildings poses particular problems. Faced with high maintenance costs owners of such buildings may be tempted to replace them with something less costly to run. Even refurbishment to improve the comfort and convenience of the building may impair its historic and architectural features. Again there is a discrepancy between the value placed on the building by its owners and the value placed on it by others, by the community. In a free market preservation of culturally important buildings would depend entirely on finding buyers willing to pay a high enough price to outbid other users seeking to redevelop the site. Because the cost of acquisition and upkeep is so high, few buildings 'at risk' could be expected to survive.

If new uses can be found for buildings no longer profitable in their existing use, they may once again become economically viable. State intervention in the form of taxes or subsidies can be used to support the change of use. But not all historic buildings are adaptable to new uses. Organizations like the National Trust play an important role by acquiring these properties. Alternatively they could be purchased by the state. The public are no longer outsiders. As taxpayers, or by their voluntary contributions, they can express the value they place on a building through the market-place. They become, indirectly, the owners.

Intervention can also take the form of administrative action – a preservation order to prohibit demolition. Historic buildings which are 'listed' cannot be altered in ways that would damage the character of the property. Unfortunately preservation orders cannot compel the upkeep of buildings. Wilful neglect is sometimes used to circumvent the listed status. Once the building becomes dangerous it can be demolished and the site is freed. More often owners do not have enough money to undertake necessary repairs. If we wish to keep our heritage we must be prepared to pay for it, whether by voluntary contributions or through taxation.

APPENDIX: SITE APPRAISAL BY RESIDUAL VALUATION

The site, which is available with planning permission for office development, is in a prime location on the edge of the central business district. Recent lettings of offices in the area have fetched rents of about £80 per sq. m after deduction of operating costs.

The plans are for a 2200 sq. m development which will give 1800 sq. m of lettable space. Allowing six months for the design and tendering stages, twelve months for construction and a further six months to fill the tenancies and find a buyer for the development means the project will need to be financed over a two-year period.

Money can be borrowed at 10%. Building costs are estimated at £560 per sq. m and the developer is looking for a profit of 10%. Investors are expecting returns of 6% on capital for property of this nature.

Gross development value

The value of the project is derived from the income it produces. If a return of 6% is required, the capital cost of the investment must not exceed 16.6 times net earnings. This figure, the capitalization factor, is the reciprocal of the target yield, i.e. 1 ÷ 6%=100 ÷ 6=16.6. It is also known as the year's purchase in perpetuity. If the cost of the project exceeds 16.6 × net earnings, the yield will be less than 6%.

Estimated rental income (net)	£144 000
capitalization at 6% = × 16.66	
Estimated GDV	£2 400 000

Development costs

These include construction work, consultants' fees and the finance costs. Not all monies are paid at the outset, so interest is only incurred as the money is borrowed. If construction payments are staged evenly throughout the time on site, on average only half the total sum will be outstanding. Professional fees tend to occur earlier in the contract, hence finance will be needed for longer, say two-thirds of the contract period.

	£
Construction cost	1 232 000
interest charges	
(10% for 18 months, × $\frac{1}{2}$)	92 400
Professional fees	
(12% of construction cost)	147 840
interest charges	
(10% for 18 months, × $\frac{2}{3}$)	9 856
Contingency	
(5% on costs)	74 105
Agent's fees for letting	
(10% of rents)	14 400
Agent's fees for sale	
(3% of GDV)	72 000
Developer's profit	
(10% of GDV)	240 000
Total development costs	£1 882 601

Residual for land purchase

The sum remaining when the development costs are deducted from the GDV must pay for the site, the finance costs and the legal and other costs of purchase. At 10% interest, over a two-year period, interest charges will total 21%. If acquisition costs add another 4% (including interest), the residual monies have to cover a sum equal to 125% of the site value.

$$\text{Residual} = £2\,400\,000 - £1\,882\,601$$
$$= £\,517\,399$$
$$\text{Land value} = \frac{£\,517\,399}{125} \times 100$$
$$= £\,413\,919$$

If the developer were to pay more than £413 900 he would be cutting into the profit margin. If the land can be acquired for a smaller sum, the potential profit is increased.

CASE STUDIES

A.

Edinburgh, whose history and buildings make it a major tourist centre, is a city of extremes. And at the moment it could be shifting from one extreme to another.

The trouble about the office market, reflected Peter Coupe, the property manager at Scottish Provident Institution, is that 'there is not a slow release of supply. We've had a glut, then a famine. Now there is a huge supply coming forward.'

There was little new development in the city centre for 20 years, so it is hardly surprising that rising demand has created an office famine. Inevitably rents have been pushed up to around £16 per sq. ft. for the best space ... Tight supply has meant that yields have been steady at around 5% in favoured central areas like Charlotte Square and around 6% further out.

Edinburgh's rise in values is not uncommon. It happened in the City of London. It is happening in other regional centres, like Manchester, Leeds, Bristol and Birmingham. And the response is the same: developers become interested in the market again. What was not financially feasible at rents of around £7 a sq. ft. becomes financially attractive at rents of £12 upwards.

'After two decades there was a strong pent-up demand for offices. We discovered that the rest of the world had caught up on tourism. So we needed offices and hotels,' says Mr Kerevan, chairman of the council's economic development and estates committee. But, he added, 'if Edinburgh is to survive, it has to have something to sell, so it has to retain its character. There is a balance between character and modern needs.'

B.

One of the more bizarre recent events in Britain's architectural and conservation history was a campaign which raged for a few weeks to obtain the listing of Alexander Fleming House, the extraordinarily ugly and forbidding headquarters of the Department of Health and Social Security at London's Elephant and Castle. Its architect was Erno Goldfinger, who provided us with concrete tower blocks galore. Steeped in the false Messianism of Modernism, he liked to say that 'to conceal a column is a crime.'

His latter-day champions insisted it was a 'work of art' but over 1000 DHSS employees say the building itself is a crime. The existing rule-book listing criteria for preservation include such abstract considerations as 'historically interesting', 'innovative' or 'especially good examples of the work of prominent architects' but say little about user-friendliness or environmental impact.

The architects who were caught up in this found all the fuss very difficult to understand. As far as they are concerned Modernist masterpieces are no more than 'buildings generally considered to be ugly, with their exposed concrete frames, steel windows, and spandrel panels which have weathered badly, where concrete deterioration has set in and surface-applied finishes have started to fail.' Such buildings invariably have very poor thermal and sound insulation and, in many cases, they also leak through flat roof or failed expansion joints. However, Fairhursts, the Manchester-based architects, have discovered 'the structures can provide a very solid base to clip on' a new facade. This creates a new identity, a new lease of life – and most importantly, a new commercial lease as well.

C.

The Cambridge Effect

Vision Park is of Cambridge but not in it – three miles north of the city centre in the village of Histon and about a mile from the M11. The A45 is the rough border between the Cambridge City Council and the South Cambridgeshire District Council and Vision Park is in the area of the latter.

The question of planning jurisdiction is important because the District Council has been more relaxed about the nature of the space at Vision Park than the city would have been ... When the Use Classes Order was reworked in 1987, the park was given a general business classification.

Merivale Moore [the developers] thought that the initial tenants at least would be high technology companies: a natural assumption given the growth of such business around the university and the success of the Cambridge Science Park. In fact 90% of the enquiries for space came from pure office users and the first two tenants were Barclays Bank and Anglia Water Authority.

Two things seem to be happening here. The first is the spread of high technology companies at a time when Cambridge was in any case growing in importance as a regional centre, thus increasing the space demands of service companies. Second there is very little space in the city itself ...

Merivale Moore, having been able to build into a rising market, has now found that changes to the Cambridgeshire Structure Plan are working in its favour. The environment secretary has made it clear that development should be pushed out along the A45 and is prepared to see part of the Green Belt lopped off for that. Vision Park, then, finds itself close to an area of likely further growth.

D.

The following are extracts from an interview with Professor Gary Davies of Manchester Business School. He suggests high street shops are failing to attract British shoppers. Retailing should look to medieval fairs and American shopping malls for fresh inspiration.

'Shopping can be functional but it can also be an event and theatre. Medieval fairs and markets were built out of town, albeit within walking distance ...

'Shoppers want to go to big out-of-town centres, but the planning system is conspiring to stop them. There seems to be a general feeling that malls destroy the green belt, but I could take you to plenty of countryside around Manchester that would be more and not less environmentally friendly as a shopping centre ...

'I accept the view that if you suburbanize retailing then you leave a hole in the middle where the city centre used to be, but that is inevitable. Customers will get what they want and stores will find a way to locate where they are needed.

'We are seeing a desperate attempt by planners to force us back into the city centres, particularly by improving public transport, but core shoppers do not want to use public transport. They want to drive.

'Why are we insisting that people do what they don't want to?'

Interview reproduced with permission from CSW – The Property Week, *18 June 1992.*

WORKSHOP

1 Using diagrams, explain why land prices are demand determined.

2 What does it mean to describe the price offered for a piece of building land as a 'residual'?

3 Explain the terms **gross development value** and **residual land value**.
What would be the effect on the GDV of the office development, detailed in the Appendix to this chapter: Site appraisal by residual valuation, of each of the following:
a property yields rise by 1%;
b building costs rise by 3%;
c rents rise by 5%.
State whether the residual land value would rise or fall in each case.

4 Distinguish between the technical life of a building and its economic life.
Use a diagram to show the effect on the lifetime of Alexandra Fleming House (case study **B**) of:
a faster than expected deterioration of panels, concrete and surface applied finishes;
b the discovery that new techniques can provide a clip-on facade to remedy many of the defects.

5 Describe and account for the main changes that have occurred since 1945 in the patterns of land use in an urban area with which you are familiar.

6 If buildings become redundant in their current use, how should we decide whether to conserve or to redevelop them?
Suggest economic policies which could be employed to aid the conservation of historic buildings.

7* How effective are market forces in ensuring land is used to the best possible advantage? Suggest reasons why the optimum outcome is not always achieved by the market.

DISCUSSION QUESTION

- 'Planning controls are essentially negative. They cannot ensure a single good or useful building is created. We should abolish them.'

19 Financing the firm

PREVIEW

- Are money and capital the same thing?
- For what reasons might a firm seek to 'raise capital'?
- Does the reason for needing funds make any difference to the way in which they are raised?
- Is it possible for a profitable business to go bust?

A growing business, with plenty of work and no difficulty finding labour and materials, may still have problems financing jobs. Money is like the oil in an engine. Finding and managing the money does not drive the wheels of business, but it is essential lubrication. It is not a resource like land, labour or capital, but it is needed before the real resources can be put to work. Money, in this context, is often called **capital**. Capital, as a factor of production, means physical assets, but here we are concerned with the financial aspects of business.

CIRCULATION OF FUNDS IN A BUSINESS: WORKING CAPITAL

Any business needs money. It needs to buy assets (plant, premises); it needs to buy materials and labour and to pay the bills, otherwise it cannot continue production. It may also make financial investments – instead of using money to buy resources for its own business, the firm buys shares in other companies or earns interest by lending funds.

Figure 19.1 shows how money circulates in a business. It is used to finance fixed capital and working capital. **Fixed capital** is tied up in assets which are used in production but are not incorporated into the product. They are long-term assets which can be used many times. Eventually they will wear out and the business must be able to replace them, so it will need to make enough money to do this.

Working capital is the money needed to finance the process of production, from the purchase of raw materials through to final sales.

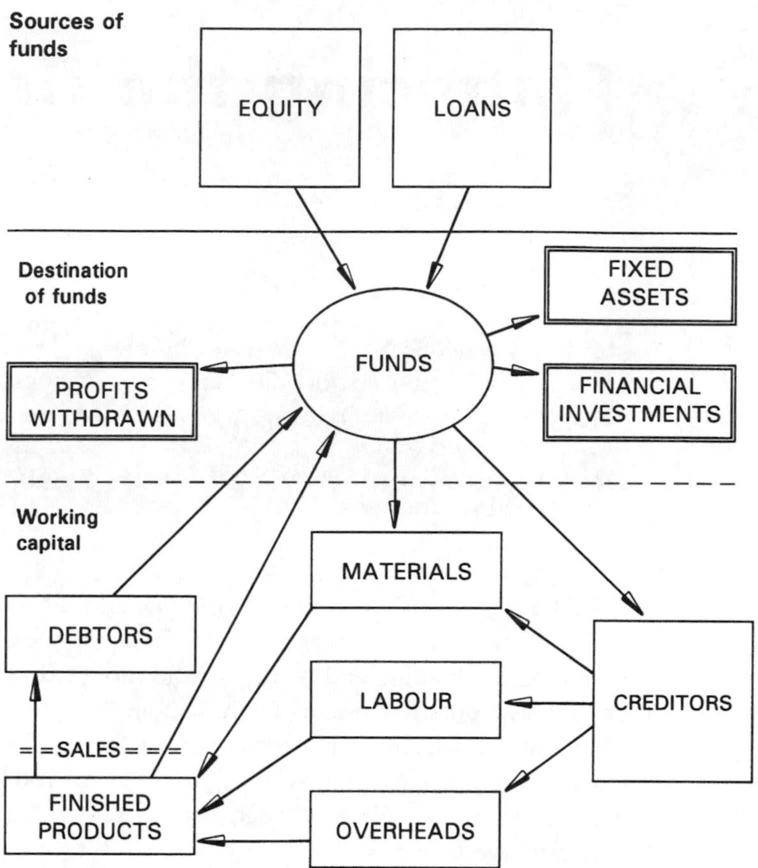

Figure 19.1 Financial flows in a business.

When the sales revenue is collected the cycle is completed and the money becomes available again. Most of it will be needed to finance continued production although a proportion of profits may be withdrawn from the business and paid out to the owners (shareholders).

To keep the business going it needs enough money to pay for all the items in the working capital sector, with the exception of 'creditors' which are delayed payments. Payments for labour, overheads and some (most) materials are usually delayed. Labour may be paid weekly or monthly in arrears, other bills monthly or even quarterly. Until then the firm need not produce the money, so the bigger the 'creditors' entry is, the smaller the amount of working capital needed. In effect the business is using its suppliers as a source of finance.

All the other items are costs. The firm must keep these as low as it can without harming its ability to carry out its work effectively. This means not holding excessive stocks of materials ('just-in-time' deliveries save money so long as they **are** in time), looking at labour requirements to avoid overmanning and keeping down the overheads, from over-used telephones to under-used plant. Holding stocks of finished goods is another cost but as the majority of building work is undertaken for a client this is not a major problem. The exception is speculative housebuilding. The final item is the box labelled 'debtors', customers who delay payments to the firm. The longer the delays the more money the firm needs to keep going, so this item also needs to be kept as small as possible.

Too little working capital is a common cause of failure in firms that have adequate sales and profit margins. Raw materials, work in progress, finished goods and payments owed by customers are all assets. The value of these assets can exceed its debts by a comfortable margin, but if it is not being turned into cash fast enough to pay the bills and satisfy the firm's creditors, they may refuse to continue supplying it. At best this is a warning signal – if the firm can improve its cash flow it will survive. At worst it is the prelude to disaster. If creditors sue and the firm is unable to raise the money, it is heading for bankruptcy.

FINANCIAL REQUIREMENTS OF DIFFERENT BUSINESSES

Initially a business is likely to be set up with a mixture of loan capital and **equity** or permanent capital. Which is preferable depends on how it intends using the funds. Fixed assets require long-term finance, while working capital only needs short-term funding. Needs vary according to the nature and size of the business. Civil engineering contractors, for example, require more funds for fixed assets than general builders. Speculative building requires a lot of working capital to finance land and materials, whereas contract work where payment is made in stages is virtually self-financing. Risks vary widely which will be reflected in the sums allowed for contingencies.

The industry as a whole requires relatively little working capital in proportion to turnover, because monies are often received before work is completed. A system of payments at intervals, as progress in the job is certified, means the client helps to pay for the work as it is done. Contractors will try to 'front load' payments on their bill of quantities, that is arrange the payment stages so they can claim money as early as possible. This, plus delayed settlements with suppliers, can produce such a favourable cash flow that working capital becomes an addition to, rather than a call upon, the contractor's reserves! Trade credit helps to pay for materials and subcontractors may be kept waiting for their money until the main contractor has in turn been paid by the client.

This shifting of the financial burden was noted in a NEDO report of 1978 as a source of concern. It is still causing friction today. Time taken to pay creditors varies considerably, reflecting the state of trade generally as well as firms' own policies. Between 1992 and 1993 the incidence of late payments to subcontractors rose from 45% to 60% and the average length of delay increased from 13 to 28 days, clear indicators of recession in the industry. With approximately 90% of the £35 bn spent on construction work going through subcontractors this represented a substantial amount of 'forced' borrowing. For the subcontractor delays also increased the risk that the main contractor might go out of business before payments had been made, possibly leading to the failure of his own business.

Subcontract labour, plant hire and an undemanding cash flow make it easy for people to set up as builders with minimal investment. Because of this too many firms lack adequate funds to secure their businesses. The result is that any delay in receiving payments, any errors in estimating or disputes over variations to the job can rapidly turn an unexpected profit into a loss. Firms then look for new contracts which will generate cash to pay the bills. But if these are won by cutting prices, risking further losses, the apparently favourable cash flow merely conceals a mounting problem. Small firms, started on a shoestring, are particularly vulnerable and help to account for the industry's high rates of start-ups and bankruptcies.

COMPANY GEARING

The permanent capital, or equity, belongs to the company. It is subscribed by its owners otherwise known as shareholders. They cannot withdraw their stake from the business, but they can sell their shares to someone else. As the value of the business grows so does the value of the equity, or share capital. Alternative sources of funds depend on some form of borrowing, either long or short term.

The extent to which a company's funding is provided by equity or by borrowing is known as its **gearing.** A highly geared company has a high proportion, 50% or more, of debt to equity. When a firm borrows it must pay interest on the loan, which raises its costs. But if, by borrowing, it can expand the business and increase profits, existing shareholders benefit from the extra revenues. Had the firm chosen to issue more shares instead of borrowing, the extra revenue would have been spread more thinly, diluting the earnings per share (Fig. 19.2). However, if borrowing favours shareholders in the boom, it penalizes them in recession. As company revenues fall highly geared firms will find interest charges take a high proportion of earnings. Lower profits depress the value of the shares.

Property companies are, by tradition, highly geared because their assets,

ALPHA PROPERTIES Debt: £700 000 Equity: 300 000 £1 shares Total funds £1m		BETA CONTRACTS Debt: £400 000 Equity: 600 000 £1 shares Total funds £1m
YEAR I Both companies achieve the same net revenues prior to interest changes. The rate of interest is 10%.		
Alpha Properties		Beta Contracts
£100 000	Revenue	**£100 000**
£ 70 000	Interest payments	**£ 40 000**
£ 30 000	Pre-tax profits	**£ 60 000**
10p	Earnings per share (EPS)	**10p**
YEAR 2 Both companies increase their turnover and their revenues rise by 40%. Interest rates are unchanged.		
£140 000	Revenue	**£140 000**
£ 70 000	Interest payments	**£ 40 000**
£ 70 000	Pre-tax profits	**£100 000**
23.3p	EPS	**16.6p**
YEAR 3 There is a slump in demand for property/contracts and revenues are halved. Interest rates are unchanged.		
£70 000	Revenue	**£70 000**
£70 000	Interest payments	**£40 000**
—	Pre-tax profits	**£30 000**
—	EPS	**5p**

Figure 19.2 Effect of gearing ratios on earnings per share.

mainly land and developments, were seen as ideal security for the lenders. Upwards-only rent reviews have helped support earnings from property. A combination of rising incomes and property values led to highly geared companies which offered shareholders exceptionally good rates of return – until the downturn in the market. With space remaining unlet, high borrowing costs became an insupportable burden and many of the high-flying property companies of the 1980s failed. Survivors worked to bring down their gearing ratios, selling assets and raising extra share capital to reduce their debts and stem the outflow of interest payments. Contracting companies usually have a much lower gearing. This reflects both their lack of suitable assets to offer as collateral and the greater uncertainties of their business. The booms and slumps of the construction world make long-term loans unattractive. The commitment to large sums in interest payments would add another risk in an already high-risk business.

SOURCES OF LONG-TERM FUNDING

Permanent capital is initially provided by the owners of a business, its shareholders if it is a company. Once established, the most important source of finance is internal – the firm's own profits. Some profit will be distributed as dividends, the rest retained for use in the business to improve its earning potential. Shareholders benefit from this investment as the rising value of the company is reflected in its share prices. How much profit is retained and how much distributed is a matter of policy. Some companies prefer to keep dividends stable, so as to preserve shareholders' income. When profits are low this entails using cash reserves, something which cannot be repeated too often or the business will suffer. Others, especially family firms, aim to plough back most of the profits.

Debentures, which can be traded in the same way as shares, are bonds (i.e. debt) issued to the public. They carry a fixed rate of interest instead of paying dividends. Unlike dividends, which are payable at the board's discretion, interest payments must be made whether the business is trading profitably or not. They are not widely used in the private sector.

Government finance is available in many forms, including tax relief, loans and grants. Two bodies were set up specifically to aid investment in industry, the National Enterprise Board (NEB), mostly concerned with larger companies, and the Industrial and Commercial Finance Corporation (ICFC) to assist medium and smaller firms. ICFC was later reorganized as a part of the Investors in Industry (3i) group and funded by the banks. A number of other venture capital funds have been set up, but despite this the 'equity gap' persists. It is a problem of finding investors willing to put up relatively small amounts of money for businesses with financial needs which have outgrown the personal resources of their owners, but which are not yet big enough to interest institutional funds. While it may be hard to find investors to bridge the gap, it is also true that many smaller firms are reluctant to surrender an equity interest to outsiders.

Sums of less than about £500 000 are difficult to find and expensive to arrange. Venture capital funds willing to provide this level of funding for a new business over a period of three to five years would look for a profit on their investment of up to 30% p.a., based on income from dividends plus the gain in share price when they came to sell. For a small enterprise this is too much. Recent initiatives have recognized the problem. The Midland Bank's Regional Enterprise Funds are one example; 3i is another group prepared to look at smaller sums. Really small firms, looking for hundreds of pounds rather than tens of thousands, may find local authorities more helpful, as some of these have also started to offer incentives to attract businesses, although their funds are very limited.

Mortgages, a familiar form of borrowing for private individuals, were also a popular commercial source of funds in the 1960s. Lenders were institutions, such as insurance companies, and interest rates were usually fixed. Inflation and soaring interest rates in the 1970s brought about a decline in this system. Mortgage debentures give the lender the added security of a charge against other assets of the company in addition to its real property holding. Lower inflation in the late 1980s brought fixed interest mortgages back into being, but the collapse in property prices at the end of the decade has left lenders wary of lending too much against bricks and mortar. Barclays Bank alone made provisions against bad debt of over £2.5 bn in 1992, much of which had been secured against property.

Property developers of prime quality commercial developments have evolved sale and leaseback arrangements. Essentially the developer sells the freehold to a long-term investor, such as a pension fund, and then takes a lease on the site for a specified term. The developer obtains a capital sum from the sale to finance future work, and an income from subletting the property. The investor receives an income from the lease and the prospect of capital gain on the freehold. This sort of arrangement was made very attractive by rising inflation in the 1970s and 1980s but has been badly hit by the recession.

Where a developer does not wish to keep an interest in the property after completion, forward sale can be arranged. As the value of the property depends on it attracting tenants who are willing to pay the estimated rents, this transfers risk from the developer to the funding organization. They will therefore seek to negotiate some form of guarantee, or penalty, against the risk that the expected rental income fails to materialize. Either the developer stakes his profit against any shortfall in rents for a specified period, say 2–4 years, or the funding organization is assured priority claim on returns from the development to give a guaranteed yield on their investment. If lettings exceed this level the extra may be shared with the developer giving increased profit. The priority yield arrangement is particularly suited to projects where costs and/or revenues are difficult to estimate. It encourages closer collaboration between funder and developer since both stand to gain from quick and successful completion. Once again the popularity of schemes like this depend on a buoyant property market.

Cooperation between developer and funding organization on single projects may provide the basis for a more settled arrangement whereby the investor, usually a merchant bank, provides general funding for the developer. The terms negotiated vary, but often include limits on the amount that can be devoted to any one project and perhaps the right of the funder to approve projects, as well as specifying interest charges, repayment procedures, etc. Individual assets may be offered as collateral or the loan may be set as a charge against the assets in general.

SHORT-TERM FUNDS

Short-term needs and smaller sums are raised from different sources to those providing substantial, longer-term funds. One of the commonest sources are trade suppliers. High interest rates in the 1970s and 1980s encouraged firms at all stages of the production process to delay payments as long as possible, since this represented interest-free borrowing. Many small builders have survived on the credit provided by their builders' merchant, while many other small subcontractors have faced disaster because of the delays in payments from main contractors. Nearly always it is the larger firm that obtains the better terms. It is more valued as a customer and creditors are less willing or able to enforce prompt payments or charge interest. The industry's suppliers have countered by offering substantial discounts for cash payments. In consequence trade credit can actually be quite expensive, but it is also very convenient, especially for smaller firms.

Bank overdrafts are similarly flexible and convenient and are the single most important source of funds for most smaller firms. Traditionally banks have preferred to advance money as working capital, since the production being financed will generate the cash for repayment. A request for general funding or long-term loans was less likely to receive a favourable reaction. Since deregulation created a more competitive climate amongst financial institutions banks have become more flexible and many small firms have come to regard their overdraft as a more or less permanent part of the business.

Although, in theory, banks can recall these loans on demand, in practice some warning is likely. In view of the problems of finding equity finance it is not surprising that small firms rely heavily on overdrafts. It does, however, make them very vulnerable if things go wrong. If the bank loses confidence in the firm, or if it needs to improve its own liquidity, it will demand repayment. Should the firm be unable to pay, the bank can call in the security, which is often the owner's house, so business and home are both jeopardized. There was a lot of criticism of banks for closing businesses in this way during the recession at the end of the 1980s. The lesson for the firm is two-fold: firstly, to watch the all-important cash flow; secondly that short-term funds are not appropriate to support longer-term activities. The problem, as we have seen, is finding long-term funds in relatively small amounts.

For the smaller firm seeking finance for plant and equipment, hire purchase may be the most practical solution. This is usually more expensive than a bank loan, but may be easier to obtain. Where firms decide to buy plant outright they must allow for its depreciation – that is funds should be set aside annually to cover its loss of value – so there is a sufficient sum available for replacement when the equipment is worn out. This is an allowable expense against profits for tax purposes. Because the money is not needed for investment immediately it is used to supplement the firm's working capital. The difficulty arises when it is time to replace the plant, but there is nothing in the piggy bank because it is all being used to pay for labour and materials!

Management of financial resources is a vital task, often neglected by small businesses because immediate problems of handling customers, dealing with suppliers, organizing subcontractors, etc., seem more urgent. Consequently many businesses that are fundamentally sound have failed because of cash flow problems. An entrepreneur who does not have the time or interest to monitor his cash flow and chase up debts can use a factor service. For a fee the factor will take charge of invoices, collect payments and even insure against bad debts, while providing his client with the bulk of monies owing to him in advance. It is an expensive service, but it can help to maintain the firm's liquidity and give the entrepreneur more freedom to concentrate on developing his market.

RECENT TRENDS

Changes in the economic climate affect the cost and availability of credit and the methods of raising funds. Financial deregulation combined with rapid economic growth in the mid 1980s produced a huge expansion of credit facilities. Building societies, who previously lent almost exclusively to homebuyers, entered the commercial sector. New instruments of credit appeared and property developers found no shortage of willing lenders as long as property prices kept rising. Bank lending to property companies rose from £2 bn in 1979 to £11 bn in 1987 and then escalated to £40 bn at the height of the boom in 1989. Large as it is this sum undervalues the banks' dependence on property prices because so many loans outside the industry were also secured against property.

As the boom evaporated, taking with it considerable numbers of developers and builders, lenders sought to limit their exposure to the risks of the property cycle. In one instance a consortium of banks offering £40 m funding for a development costing £50 m required the developer to obtain insurance against loss for the whole of the sum borrowed, thereby shifting the risk from the lenders to the insurers. The fall in property prices left a legacy of bad debts which is likely to hamper banks and borrowers in financing recovery.

The recession exposed a number of weaknesses in the way in which British industry obtains its funds. It has drawn attention to the equity gap for smaller firms and to the folly of regarding property as risk-free collateral. It has also emphasized the importance of good management and sound financial controls. Too many people enter business without any understanding of budgeting and cash flow. A survey amongst insolvency practitioners in 1993 blamed poor management for a third of company failures and lack of working capital for almost a fifth. Less than one in twenty failures were ascribed to loss of long-term finance – in other words the banks.

The banks have lessons to learn just the same. They relied too heavily on collateral and did not pay enough attention to assessing the risks attached to individual borrowers. The immediate danger is that banks will be over-cautious in avoiding loans to property-related businesses. As signs of recovery emerged in 1993 the Building Employers Confederation expressed concern at banks' unwillingness to support firms who needed to increase borrowing to finance new work. Banking in the 1970s and 1980s became more centralized. Lending policies were decided at head office, leaving branch managers with little discretion over how much and to whom they should lend. A return to more responsibility at local level coupled with greater emphasis on knowledge and assessment of the borrower's business would allow for more flexible lending to suit firms' needs.

The big four clearing banks are the dominant source of finance for business. More than half of company borrowing comes from them. Traditionally banks have been lenders not investors but changes in banking practices may be necessary to develop the sort of long-term financial support which is crucial to corporate health. Innovation in finance reflects market changes in the same way as innovation in other areas. What remains unchanged are the fundamentals – the need to ensure adequate cash flow, the ideal of matching the timescale of finance to asset structure and the need to assess risks thoroughly.

CASE STUDIES

A. Little and ...

Winchester based business man David Meadham is an entrepreneur of the Thatcher era. In 1977 he set up his own plant hire company when a builder he was working with could not afford the repair bills for a mixer. His previous experience with Bath Plant Hire helped him prosper and within a few years he had opened additional hire centres in Andover, Basingstoke, Bournemouth and Portsmouth. Last year turnover reached £4.5 m. So why is he auctioning 45 excavators, 56 dumpers, 40 rollers, assorted vehicles, forklift trucks, compressors, pumps and other equipment – in all some 15% of his plant?

Mr Meadham explained, 'In the last nine months we have had more bankruptcies and debts than I have seen in my thirteen years of trading as a company.' The plant being auctioned represents half the equipment that he has had in his yards instead of out on hire over the past few months. 'People who have heard of the auction have telephoned to ask if I have called in the receivers. It's because I don't want that to happen that I'm taking this action now.'

Based on a report in the Southern Evening Echo, *6 September 1990, reproduced with permission.*

B. ... large

Mr Alastair Morton, chief executive of Eurotunnel, was in a sombre mood this weekend after his bankers said they were having trouble raising the extra £2 bn of loans which the project needs. Together with a proposed £500 m rights issue this autumn, the money is vital to cover its soaring costs.

Eurotunnel's loan agreement with the syndicate of 210 banks requires it to have enough finance to cover the entire cost of the project. At present it has £6 bn with costs estimated at £2.5 bn; if the banks fail to raise the funds they will prevent Eurotunnel from meeting the banks' own requirements. At the same time Eurotunnel is restricted from seeking new lenders without the agreement of the present consortium. Underwriters for the rights issue have been found and new agreements have been reached with the contractors on costs, so there is little to be done except wait on the banks.

What are the possibilities? If no agreement can be found, the tunnel could come to a premature halt. This is unlikely at a time when the project looks more realistic than ever before; no one wants to be responsible for pulling the plug on it. The most optimistic scenario is if the banks raise the full sum. The current economic climate, especially in Japan, which is responsible for nearly a third of the syndicate, makes this unlikely. If the rules were relaxed to allow Eurotunnel to seek new lenders, it is difficult to see where these might come from. American banks are suffering from the collapse of property loans; public funds are ruled out because the project has been a private sector venture from the outset with a high-risk profile. Most likely is some amendment to the agreements to allow work to continue on an interim basis. The longer work continues, the better the tunnel looks and the more likely that the extra finance will be forthcoming.

Based on an article in Financial Times, *© F.T., 17 August 1990, reproduced with kind permission.*

C.

One of the UK's largest property companies announced today (27 May 1993) a £199 m rights issue. Hammerson intends to use the money to reduce its gearing, currently at 114% compared to 68% last year, and to give itself a breathing space in which to review its organization.

The intention is to overhaul its portfolio; some disposals are expected but there are also plans to spend half of the money raised on improving properties. At the same time the company's management will be reviewed and substantial savings in administration are promised.

The new shares, issued on a 7-for-15 basis, dilute net asset values by 14%. The issue has been welcomed by the stock market.

WORKSHOP

1 Distinguish between debentures and shares.

2 What is meant by 'gearing'?
Explain the advantages/disadvantages of raising money by loans, as against issuing additional shares in a company.

3 Why are property developers generally more highly geared than contractors?

4 Consider case studies **A** and **B**. What alternative courses of action could have been taken by the plant hire firm? Consider the pros and cons of each.
Why do you think Eurotunnel did not try to increase the sum it was intending to raise from shareholders when the banks indicated they might not be able to provide as much loan capital as expected?

DISCUSSION QUESTION

'Delaying payments as long as possible is just good commercial sense.' 'Late payments put people out of business.'

❑ Should firms who are slow to pay be penalized? If so, how?

20 Externalities and cost-benefit analysis

PREVIEW

- Why are some things, like roads or hospitals, paid for out of taxes while others, like food or housing, are priced in the market and paid for by consumers?
- What costs arise from the decision to journey to work by car and who pays them?
- How can we decide whether a project, such as a new road, is worthwhile if there are no prices for its use to show how much people are prepared to pay for it?
- What is the difference between financial appraisal of a project and a cost-benefit analysis of it?
- Can the price mechanism allocate resources efficiently where there are costs/benefits to people other than the purchasers of the goods produced?

EXTERNALITIES

In a market economy consumers and producers make their decisions on the basis of price. Consumers buy at the lowest price they can find, while producers sell at the highest price they can achieve. Competition keeps prices down and forces producers to use resources efficiently. The goods which are produced are the ones consumers want, as indicated by their willingness to pay. Individual interests through the magic of the market-place, or what Adam Smith called 'the invisible hand', result in maximum benefit for all.

Of course the perfect market is a model, an abstraction from reality, but some important points emerge from it. Although, in the real world, consumers do not always seek the lowest price, they do look for good value. They recognize goods vary in quality, but they will prefer a low price to a high price as long as the goods satisfy them in terms of quality, style, etc. Producers are not always as efficient as the model suggests and they may form cartels to try to raise prices. Despite these qualifications competitive markets generally reward efficiency and encourage producers to match consumer requirements.

The model provides a basic guide to consumer and producer behaviour in a simple economy, but even in Adam Smith's day it was recognized that not all needs could be met in this way. Some goods could not be supplied by the market-place. The most important of these was the nation's defence, an example of what economists call **public goods**. Public goods cannot be supplied to one consumer without becoming available to all, they are **non-exclusive**. If a country's defence system keeps the peace, everyone shares in it – it is not possible to exclude anyone. This makes a market in peace-keeping impossible, for if peace is made available whether we pay or not, who will choose to pay? To ensure adequate provision of public goods, they must be paid for by a levy on all the 'consumers' – by taxes. Since individuals cannot choose how much security to buy for themselves, the level of defence spending is a political or administrative decision. It is not determined by market forces.

Everyone shares in public goods so it is right that decisions on them should be made by the community, but sharing does not reduce the benefit to individuals. Use of public goods is **non-rival** as well as non-exclusive. Peace for all does not lessen the peace enjoyed by each. This contrasts with individually acquired goods, such as housing. Not only can those who do not pay be excluded from housing, but the more who share a house the less space there is per person. Overcrowding reduces the utility for all the occupiers.

Many goods today are paid for wholly or partly through taxes and consumers have little direct say in the level of output. Most of these share the characteristics of public goods to some degree. Roads, for example, could be supplied through the market. People wanting to travel would have to pay the road supplier in order to use the road. Toll roads do exist but they are inconvenient and we prefer to treat most roads as public goods. But even though we make them non-exclusive we cannot make them non-rival. Rush hour travellers in any city know that as road space becomes congested with more vehicles, their own utility is diminished. (Problems of congestion are now so acute that rationing road use through some form of pricing is being considered.)

Merit goods, like public goods, are often part funded through taxation. Unlike public goods they are used on an individual basis, but individual consumers may undervalue the products and buy too little. Education, health and housing are possible cases. Many people are willing to spend a great deal of money on these, but others may rate expenditure on holidays as more important than house repairs. Society has an interest in these decisions because they have spillover effects on other people. Too little maintenance leads to decaying property and poor living conditions. Health suffers as a result. Too much time off work, or school, increases the risk of unemployment. All these problems impose a cost on the community as well as on the individual who suffers them.

Such spillover effects are termed **externalities**. They are the **public costs** and **public benefits** of decisions taken by private consumers or firms. As consumers we seek to maximize our own welfare: we weigh up the price of an item against the benefit we obtain. Of course, if we recognize that our decisions affect others, we can modify our actions accordingly. We may refuse to buy certain products because of the manner in which they are produced, or pay extra for environmentally friendly goods, but the price mechanism does not help us make these decisions. They reflect concerns outside the market-place. How much we are prepared to pay for environmentally friendly goods depends on how strongly we feel and what we can afford. Our generosity is balanced by an opportunity cost, for we have limited resources. The economic model is still relevant, prices affect decisions, but those prices fail to give a true picture of the costs or benefits when there are significant externalities present.

Because the market-place operates in this individualistic way, it often fails to make the best use of resources. Goods where part of the cost falls on others will be over-produced, because consumers do not have to pay the external costs. Where there are external benefits, these will be undervalued as consumers are only willing to pay for what they consider worthwhile. Costs and benefits to other people tend to be perceived less readily and valued less highly than those we experience ourselves. As a result socially desirable goods may be under-produced.

This is shown in Fig. 20.1. In a free market consumers will balance the benefit from, for example, electricity with the price they have to pay. Assuming that the marginal cost of producing electricity increases as output rises (see Chapter 9) consumers will be willing to pay for q^2. The production costs here are those costs borne by the electricity companies and passed on to the consumers in the price for electricity, i.e. the costs of fuel, labour, plant, etc. These are the private costs. In addition there will be some external costs, such as damage to forestry caused by acid rain as a result of emissions from power stations. If these are included, the marginal social cost (public + private costs) of more electricity is seen to be higher than the price charged to consumers.

Moreover marginal cost to society rises more steeply than the marginal cost paid by consumers, because the effects of pollution become more serious as pollution levels increase. Assuming that there are no external benefits, the demand curve which shows the benefits to consumers (MPB) will be the same as the overall benefit to society (MSB). Social welfare is maximized at q^1. Beyond this point the extra cost to society of generating more power, including the additional pollution, exceeds the extra benefits. Consumers are particularly likely to undervalue the external costs in cases like this where the damage is quite remote, perhaps even in another country. Even if they are aware of the pollution costs they have no way of setting a price on their own contribution to them and adjusting their consumption of electricity.

These problems may be tackled by making the market take fuller account of the costs/benefits involved. A tax on pollution will lead to higher prices. Producers can choose to use the extra sales revenue either to pay the tax or to finance investment in equipment which will reduce emissions and so avoid the tax. If the tax revenues are also used by government to clean up the effects of pollution the higher prices reflect the benefits of a cleaner environment. Consumers now pay a price which covers the full cost of production, including making good or preventing the polluting side-effects. As they adjust to the higher prices demand will contract because they try to economize in their use of the goods. If the tax is set so as to raise the market price to the level of the marginal social cost, the market can achieve the optimal position q^1. The external cost has been internalized and resources are reallocated accordingly (Fig. 20.2).

The idea, advocated by environmentalists, of a carbon tax is intended to work in this way. A tax on carbon-based fuels serves the dual function of

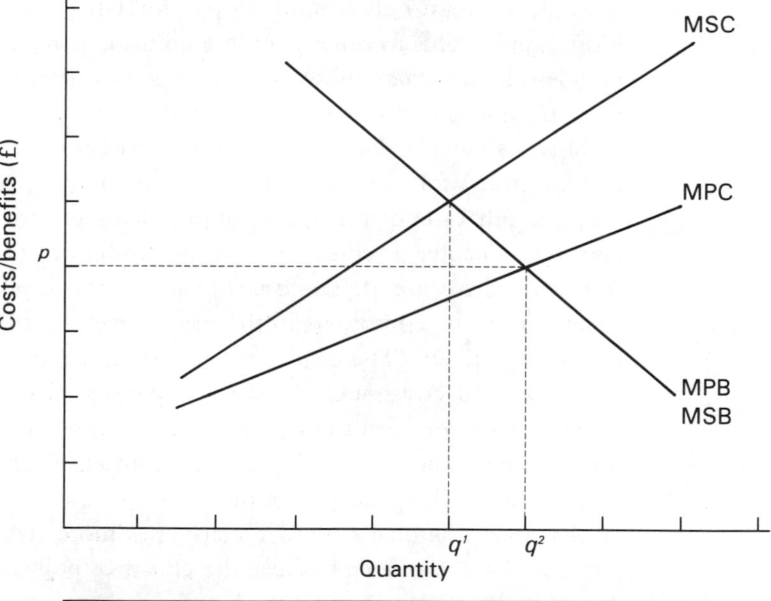

MPB – marginal private benefit of additional units to consumers.
MPC – marginal private cost of supplying extra units.
MSC – marginal social costs: the divergence between MSC and MPC is due to external costs arising from the product.
MSB – marginal social benefit: assuming there are no external benefits MSB = MPB.
p – market price.
q^1 – output which maximizes social welfare (MSC = MSB).
q^2 – equilibrium quantity in a free market (MPC = MPB).

Figure 20.1
External costs result in overconsumption.

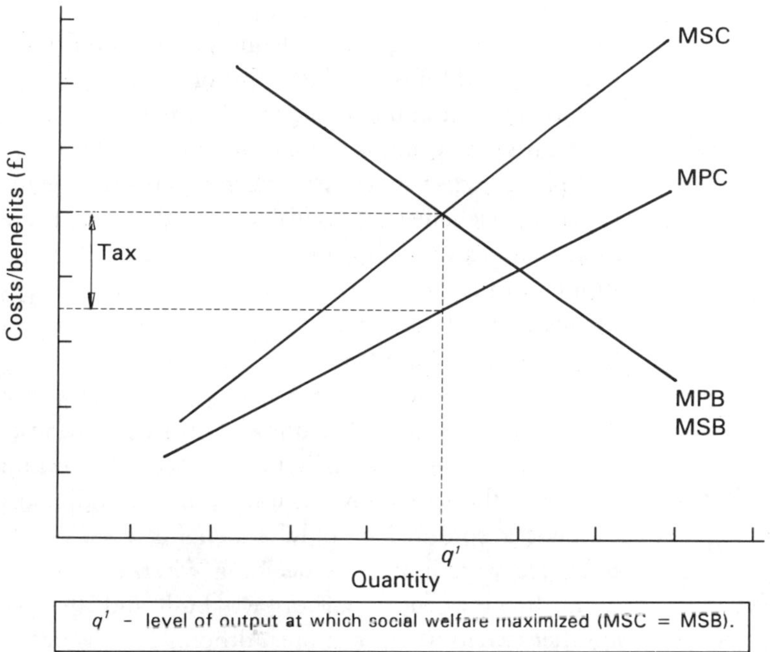

Figure 20.2 Use of taxes to internalize external costs.

making us economize on the use of these fossil fuels, which are non-renewable, and encouraging the development of alternative technologies which will be less polluting. It makes us value our resources more carefully.

The same principle lies behind the idea of road pricing, a policy to encourage better use of road space by making drivers more aware of the true cost of their journeys. A vehicle entering a heavily used road system adds to existing congestion and causes even longer delays. Journey costs are marginally increased for everyone using the road. Where there is heavy traffic the true (social) cost of one additional journey is much greater than the individual road user realizes.

Various ways of pricing scarce road space have been suggested. One of the most flexible is an automatic metering system installed in vehicles and triggered by sensors in the road. Charges could be varied with time and place, to reflect the high cost of congestion, which is mainly a problem of urban areas in peak hours. Raising charges in these circumstances would help to ease the traffic jams as some expensive, but less essential, journeys would be rerouted or rescheduled – or even abandoned.

COST-BENEFIT ANALYSIS

The price mechanism cannot be made to work in every circumstance. We may not wish consumers to pay a full economic price to use public sector facilities such as roads, schools or hospitals, but providing these facilities uses scarce resources. Without prices and profits, how can we judge whether those resources are being put to good use, or whether they might be better used for some alternative project?

A useful tool in public sector investment appraisal is **cost-benefit analysis (CBA)**. The aim is simple: to evaluate all the relevant costs and benefits (including externalities) to determine the net gain to the community. If alternative schemes are available, e.g. different routes for a proposed road link, these can be compared to ascertain the best option. In principle it is simple; in practice it can be a complex, demanding and expensive exercise. Hence it is only used when major projects, such as an airport or motorway link, are under consideration.

The first step in any CBA is to define its objective, which should be to find the most beneficial solution. This might mean taking no further action – leaving the situation as it is. The Roskill Commission (see Table 20.1) was given the task of investigating a third London airport. (Strictly speaking this was a cost analysis only as benefits were not evaluated.) Its terms of reference were limited to deciding when and where, but not whether, to build. In the event no airport was built and the need did not reappear for another twenty years or more. Although it was not a full CBA, the meticulous approach of the Roskill report shows the strengths and problems of this type of appraisal.

The question of timing was essentially an analysis of opportunity costs, comparing the benefit of using resources to build the airport with leaving them available for other uses. By building sooner they would minimize the costs of mounting congestion at London's existing airports. By building later they would save the costs of construction for another year. They calculated the money needed to build the airport was capable of earning interest of £22 m per annum. Each year's delay 'saved' £22 m, but each year the costs of congestion increased. The project should be timed to equalize marginal cost and marginal benefit – the new airport should therefore be built by 1982 when they estimated congestion costs would have reached £22 m p.a.

Deciding the best location was a matter of comparing costs between sites. Construction costs did not differ greatly, but what costs would be incurred by the community around the airport? Defining a relevant cost is not always simple. Increased road traffic generated by the airport would necessitate building new road links, clearly a significant and relevant item. Traffic effects may spread over a considerable distance; at what point do they cease to matter? Similar decisions had to be made with noise and other environmental effects. Costs are not purely financial – any loss of welfare is a cost to be set against the benefit which the airport is presumed to bring. Environmental impact might include disturbance to wildlife – is this to be included? Housing on the airport's perimeter will suffer substantial loss of amenity, lowering its value; houses further away will suffer less severely. A

Table 20.1 Comparison of inter-site costs for third London airport (based on estimates made by the Roskill Commission)[1]

Cost item	Cublington	Foulness	Nut'stead	Thurleigh
	(£m discounted to 1982 values)			
Expanding Luton airport	0	18	0	0
Airport construction	18	32	14	0
Airport services*	23	0	17	7
Weather conditions	5	0	2	1
Flying times*	0	7	35	30
Passenger access costs*	0	207	41	39
Freight access costs*	0	14	5	1
Road investment	0	4	4	5
Rail investment	3	26	12	0
Air safety (bird strikes)	0	2	0	0
Closure or relocation of:				
Defence establishments	20	0	5	61
Govt. research stations	1	0	21	27
Private airfields	7	0	13	15
Disturbance (inc. noise) to:				
Residential areas	24	0	70	11
Schools	7	0	11	9
Agriculture	0	4	9	3
Commerce & industry	0	2	1	2
Recreation	13	0	7	7
Luton airport noise	0	11	0	0
Aggregate differences*	**0**	**197**	**137**	**88**

1. In the 1960s fears that Heathrow and Gatwick were reaching saturation point led to the setting up of the Roskill Commission. Its remit was to investigate how soon a new airport would be needed and 'which site should be selected'. The Commission sat for 2½ years and cost over £1 m. An initial list of 78 possible sites within an 80 mile radius of London was whittled down to four front runners. A detailed cost comparison was prepared for these. They were: Cublington (Bucks.); Foulness (Essex); Nuthampstead (Herts.); Thurleigh (Beds.).
The differences in costs between the four sites is summarized above. The lowest cost site under each item is shown as 0. For some difficult-to-cost items (*) high and low estimates are calculated. These did not affect the rankings and the figures shown are the higher series. The choice of Foulness would have entailed expanding Luton airport also; had any other site been chosen Luton would have been closed. The difference between the least cost option, Cublington, and the most expensive, Foulness, amounted to less than 4% of the total cost of the project.

line has to be drawn at some stage where the damage no longer counts.

Once the relevant 'costs' have been decided upon, they must be valued. Some are straightforward. Construction costs can be estimated at market prices, which can be taken as a good indication of the value of those resources. In other cases the market price may be distorted; land in a green-belt area has a low market value because there are planning restrictions on its use. If it is being considered as an airport site, should it still be valued as green-belt land, or as building land? What is the opportunity that is being sacrificed, open space or alternative development?

Often there is no market at all and a 'shadow' price must be set.

With transport projects a major consideration is travel times. How do you value people's time? Noise is another significant effect of airport development, but there is no simple way of putting a price on peacefulness. A fall in property values may reflect loss of amenity to some extent, but property prices are subject to many other influences. The cost of double or treble glazing provides a partial remedy to the nuisance, but does not cover the loss of amenity of open windows or peaceful enjoyment of the garden. Can we test people's willingness to pay for peace and quiet? Or how much they are willing to accept as adequate compensation for the disturbance? (The answers to these two questions are not always the same!)

The outcome of a CBA can depend on which costs/benefits are included and how they are valued, so these are crucial decisions.

The estimated lifetime of the project is the next important variable. It is economic life rather than technical life that is relevant. The timespan for major public sector projects, of the sort where CBA is likely to be used, is often in excess of 20 or 30 years, whereas much private investment appraisal is looking at a timespan of five to ten years. Even commercial buildings may not have much more – 16 years is the estimated lifespan for retail outlets in American cities. The longer the timespan involved, the greater the uncertainty. Forecasting is largely based on the projection of current trends. Over a long period these are likely to deviate considerably from the projections as circumstances change.

In the case of an airport the benefits depend largely on the projections of growth in air traffic, for holiday and business purposes. This is influenced by many variables, including the state of the economy, growth in incomes, the capacity of alternative transport systems, the price of travel, etc. The need for London's third airport, which appeared urgent in the 1960s, diminished in the 1970s. Higher oil prices put up fares and slowed the growth in demand, while bigger aircraft could carry more passengers without the need for more runway space. The further into the future estimates are projected, the less reliable they become.

The long timescale emphasizes problems in comparing costs and benefits which do not occur simultaneously. In most projects the bulk of costs are incurred early on, in the construction phases, with some outgoings continuing through the lifetime of the project. Benefits typically start later, when the scheme becomes operational, and continue to build up over a period of time. In comparing costs and benefits, the time value of money is important. One pound today, earning 10% interest, will be worth £1.10 a year from now. One pound in a year's time must be discounted by 10% to calculate its present value.

The choice of discount rate may be crucial to the outcome of a CBA, especially where a long time period is involved. A high rate places a lower value on the more distant returns and makes it harder to justify an expensive scheme, where benefits extend over a long period. Many public sector schemes are long term and are likely to suffer when compared with commercial investments which will produce much quicker benefits. The question arises whether public sector investment should be expected to show the same rate of return as private sector schemes, i.e. use the same discount rate, or should a lower rate be employed to give greater weight to the benefit the community will gain from taking a long-term view?

Adjusting the discount rate is also a simple, if rather crude, way of allowing for risk. Applying a higher discount rate to a high-risk project will mean it has to achieve bigger returns to show a net benefit. Although this is common commercial practice it tends to distort results where a long timescale is involved. A more sophisticated approach is through sensitivity analysis and the use of probability theory to test the effects of changes in particular variables in the appraisal. Calculations can be tried with optimistic figures and pessimistic figures to establish how vulnerable the project is to changes in particular elements, such as interest rates. This provides a stronger basis for decisions as it produces evidence of the likelihood of certain risks and the degree of variation from the initial estimates which can be absorbed before the overall benefit is seriously affected. Roskill used two different valuations for time saved by airport users because of the difficulty in putting a price on this, but whether the sums were done with the higher or the lower figure did not affect the outcome in this instance.

Even if the final outcome of the calculations is a satisfactory net benefit, there is a further element to consider, the distributional effects. Costs and benefits are unlikely to fall equally on all sections of the community. The main beneficiaries of a third airport for London would have been air travellers, a relatively wealthy group. The main potential losers were the households who would suffer relocation and/or nuisance, especially from noise. These were mainly lower income groups, largely agricultural workers, since the chosen sites avoided large urban centres of population.

Strictly speaking distribution of benefits falls outside the remit of the cost-benefit analysis. If welfare is unevenly distributed society can use measures such as taxes and subsidies to create a better distribution. The analysis is intended only to discover if the project is likely to improve total welfare; theoretically, if there is a net improvement gainers could compensate losers and still be better off. In practice this is unlikely to happen. Moreover, where shadow prices are based on questionnaires to test the willingness to pay, the resultant figures will give undue weight to the desires of the wealthy who can afford to pay more than the poor to realize their wishes. Decision-makers may wish to take these factors into account.

Coming to a decision is the final stage. CBA is not an automatic process – results depend on the parameters selected. It is a tool the value of which lies in the discipline it imposes on decision-makers to clarify their objectives, to define the factors relevant to the issue and to quantify the values of these as objectively as possible. The recommendation that is offered may be rejected, but any decision will have to be justified in the light of the evidence presented in the CBA. The final outcome may be influenced by other factors, such as political expediency – a beneficial project is not necessarily a popular one – or social concerns, especially where costs and benefits are unevenly distributed. If the scheme that produces the greatest net benefit is also the highest cost alternative, simple lack of funds may lead to its rejection.

The Roskill Commission concluded that the difference in cost between the most expensive and the least expensive of the airport sites they examined was likely to be less than 4% of the total costs involved. A 1% error in the projections would have reversed the results. In view of the problems of quantifying many of the factors involved, especially noise and time savings, and the lengthy timescale of the project with its inherent uncertainties, such a small margin cannot be considered decisive. Had the airport been built, it is likely that the government's preference for the most expensive site, on the political and social grounds that fewer people would suffer disturbance, would have overridden the Commission's recommendation in favour of the least expensive site.

VALUE OF COST-BENEFIT TECHNIQUES

A full-scale CBA is an expensive undertaking, but the use of similar techniques on a more restricted scale has become widespread. Local authorities with urban renewal projects have found this approach helpful in comparing options, ranging from complete clearance and redevelopment at one end of the scale, to refurbishment of existing property at the other.

An imaginative housing improvement scheme in Jerusalem was based on adding a new section to the front of existing blocks, which increased the amount of living space, improved circulation patterns and transformed the facade. All this was achieved without moving out tenants and without breaking up the existing communities. The estate enjoyed a face-lift which improved morale and lowered vandalism in the area. Benefits like these can be entered into appraisals of urban renewal schemes, alongside the costs of construction and direct financial gains from higher rents or rates.

An important part of CBA has been its attempts to grapple with the problems of valuing costs or benefits which do not have market prices. Such things as clean air, clean rivers, open countryside and unpolluted seas lack prices because there are few clearly defined property rights attached to them, they cannot be bought and sold. This does not mean that we do not value them. The environmental movement shows just how greatly we do value these amenities – and other parts of our natural heritage which we may never experience personally, such as the tropical rain forests or Antarctic wilderness. CBA has helped to develop techniques based on willingness to pay to allow us to enter these benefits into the balance sheet against the more obvious gains from production and development. This way we can move towards more rational decisions on the use of resources, weighing costs against benefits.

CBA is undertaken from the viewpoint of the welfare of the whole community, whereas ordinary investment appraisal measures only the welfare, in the restricted sense of financial gain, of a part of the community. Since public authorities seek to operate in the interests of us all, the wider type of analysis is suited to the public sector. As environmental concerns gain greater currency we may see private sector firms adopting similar techniques, especially in relation to major projects involving construction. The built environment is something which affects the quality of all our lives. A summary of the main approaches likely to be encountered is given below.

- **Cost-benefit analysis:** quantifies relevant cost and benefits and evaluates policy against the net gain.
  – *Advantages:* prices non-market costs and benefits in terms of value to consumer (willingness to pay); provides a way of measuring economic efficiency of alternative outcomes.
  – *Disadvantages:* expensive to gather necessary information; fails to take into account distribution of costs/benefits; may omit difficult-to-measure items.
- **Cost-effective analysis** (Roskill Commission): compares cost of different schemes to achieve a given outcome.
  – *Advantages:* less information to handle; concentrates on more easily measured items.
  – *Disadvantages:* no comparison of costs against benefits; recommendations not necessarily the most economically efficient outcome.
- **Environmental impact assessment:** Attempts to identify and measure total impact, good and bad, of a policy on the environment.
  – *Advantages:* wide ranging; may incorporate standard CBA; not restricted to monetary measures of costs/benefits.
  – *Disadvantages:* difficult to use; criteria not clearly defined; does not suggest 'best buy'.

- **Risk-benefit analysis:** Compares the benefits of a course of action with the risks incurred, e.g. the benefit of building a nuclear power station against the likelihood and scale of damage from nuclear accidents.
  – *Advantages:* focuses attention on degree of risk; compatible with other forms of CBA.
  – *Disadvantages:* does not itself lead to a decision.

CASE STUDIES

A.

Consider the Department of Transport figures given in Table 20.2 for the numbers of motor vehicles licensed in the UK 1961–1987. Assuming steady economic growth, forecasts for the increase in traffic by the year 2025 range from 80% to 140% higher than in 1987.

B. Too little road, too many cars

As far back as the 1960s a Danish traffic engineer, Professor Bendtson, observed a correlation between the speed and use of roads. When pressure of use reduced speeds to around 5 mph the volume of traffic ceased to expand so fast, as drivers sought to avoid the delays.

Journey time and costs rise in an exponential progression as more vehicles lead to more congestion. It has been calculated that an additional vehicle entering a traffic system which is flowing at 20 mph causes 10 seconds delay to other traffic for each minute it is on the road. At 10 mph the delay increases to 2 minutes. At 5 mph, Bendtson's minimum workable speed, the delay imposed by an extra vehicle rises to 10 minutes.

Despite this Bendtson did not advocate road building to solve congestion . . .

Table 20.2 Motor vehicles with current licences (GB)

Vehicle type	**1961**	**1971**	**1981**	**1991**
Private cars	5 979	12 085	15 392	21 952*
Motorcycles	1 869	1 124	1 459	750
Public transport	92	106	110	109
Goods vehicles	1 457	1 632	1 763	449*
Others	510	531	631	1 951
ALL	**9 907**	**15 478**	**19 355**	**25 211**

*Since 1990 light goods vehicles (under 3500 kg gross weight) grouped with private cars.

C.

Bankok driven potty as traffic jams grow longer

Don't leave home without one – a potty, that is – drivers in Bangkok have been warned by the Thai government. They should also take emergency supplies of food and water. This alarmist advice follows a week in which Thailand's capital, already famous for its traffic jams, excelled itself.

A series of nightmare snarl-ups paralysed the city for hours, stranded some children at school until after midnight and delayed international flights, prompting the government to advise drivers to carry rations and chamber pots – just in case.

Commuters took up to six hours to drive a few kilometres; on most routes there is no alternative to road transport . . .

Bangkokians have developed a peculiar culture to deal with the physical and mental strain of their predicament. Even people of relatively modest means employ a driver and use the back seat of the car as a secondary office, complete with newspapers, a mobile telephone and sometimes a television.

An invitation to lunch is followed by a period of delicate negotiation. Each party attempts to name a restaurant close to his or her office, and the loser is permitted to be late and to boast of the appalling traffic jam which trapped him.

Traffic lights often add to the misery. Many are operated by hand, and one set has been known to take 18 minutes to change . . .

Bangkok's traffic congestion threatens to choke the country's dynamic economy. Some expatriates have moved to the efficient and less polluted island of Singapore rather than struggle to reach meetings in Bangkok.

D. The channel tunnel

In 1986 Britain and France gave the go-ahead for a major new cross channel link, under the sea. The tunnel, which was due to open in 1993, was expected to cost £4.5 bn and its operators, a consortium called Eurotunnel, were given a 50 year franchise. By 1990 the estimated costs had escalated to £8.5 bn and for a while it seemed that the project might founder, until arguments between Eurotunnel and the construction companies as to who was responsible for the over-runs were resolved and additional finance was raised. In the early stages of work the geological formations proved less homogeneous than expected and progress was initially slow.

From the outset the tunnel has provoked controversy. As 'green' issues gain attention, the ecological impact of the tunnel has raised hackles, especially among those who live close by the terminal area. Disposal of the spoil in a manner that would not cause undue environmental damage was one problem. The bulk of the estimated 4.7 m cubic metres of material brought out is being dumped close to where the tunnel emerges in a land reclamation scheme.

Those in favour of the tunnel point to the jobs created by the construction work, nearly

Table 20.3
Eurotunnel estimates of passenger and freight demand

	1985	1993	2003
Passengers (million trips p.a.)			
All transport systems	48.1	67.1	93.6
By tunnel	—	29.7	39.5
Freight (million tonnes p.a.)			
All transport systems	60.4	84.4	122.6
By tunnel	—	14.8	21.1

4000 workers, and in the orders for other industries for cement, engineering equipment, etc. Later there will be locomotives and carriages to build and when it is operational it is estimated 3000 workers will be needed to run services and maintain the tunnel. Those who are less enthusiastic query how many of the jobs will be won by British firms. They point out there are also potential job losses, perhaps as many, or even more, in the ferry ports if a large share of the cross channel traffic diverts to the tunnel.

The regional impact of the tunnel is likely to favour the South East as businesses are attracted to Kent and easy access to Europe. Much depends on the road and rail links to other parts of the country. Liverpool foresees potential benefits if goods could be quickly routed across the country and into Europe. Atlantic vessels might then use Liverpool as a major entrepot for European–American trade. South coast ferry operators are unlikely to lose a great deal of bulk freight but are concerned about the impact on passenger traffic and the higher value freight in container lorries for which they have invested heavily in ro-ro ferries.

Road hauliers and British Rail are also going to be affected. Much depends on the total volume of traffic (Table 20.3). A high rate of economic growth will allow the extra capacity offered by the tunnel to be employed without detriment to other carriers, but if trade does not expand as fast as expected the competition for customers is likely to be fierce. At present nearly 80% of goods bound for Europe are carried by road. Will the tunnel enable British Rail to win back some of this trade?

WORKSHOP

1 Consider the data in case study **A**. What is the basis for the Department of Transport's traffic projections, and what are the implications of the figures?

2 ***a*** Case study **B** states each vehicle, in a congested road system, imposes a 2 minute delay on traffic when speeds are down to 10 mph. Assume fuel consumption at this speed of 10 mpg, fuel price of £2 per gallon, and 250 vehicles affected: what is the marginal public cost, for extra fuel consumption, of 1 extra vehicle?

b The Confederation of British Industry estimated the costs to industry of traffic congestion at £15 bn. What are the main elements of these costs?

3 Can you suggest other, less easily measured, public costs which may result from traffic congestion (see case study **C**)?

4 Professor Bendtson in case study **B** did not believe road building was the solution to traffic congestion. Why not?

5 The alternative to more roads is fewer cars. Outline different ways in which car use could be restricted. Which measures do you think the best?

6 List as many costs and benefits as you can that ought to be included in a cost-benefit analysis of the Channel Tunnel (case study **D**). Are there any significant items which do not have a market price? How would you try to value these?

DISCUSSION QUESTIONS

- ❑ How important are the traffic estimates in the analysis referred to in question 6? Discuss the problems encountered in making these estimates and how they affect the outcome.
- ❑ A financial appraisal was undertaken of the tunnel scheme, but not a CBA. Should there have been a proper CBA? Explain your answer.

21 The national income

PREVIEW

- How can we judge the performance of the whole economy?
- What are the links between spending and income?
- What will happen if we decide to spend less and save more?
- How does consumption affect investment?
- Why does the economy suffer from cycles of boom and slump?

So far we have been dealing with economic decisions largely at the level of individuals, either as firms or consumers. We turn now to the whole economy and the ways in which government can influence economic performance.

MEASURING ECONOMIC PERFORMANCE

Nations, like individuals, generally measure their economic performance in terms of their annual income. National income is the aggregated earnings of all the enterprises and individuals within the society. Since income is earned by producing and selling goods we can equally well use output or expenditure to measure the economy. Theoretically all three measures should be equal: if a car is bought for £10 000 we have **expenditure** of £10 000 by the purchasers in exchange for **production** which is worth £10 000, and the producer receives **income** of £10 000 which is shared out among the workers, shareholders and suppliers who contributed to production (Fig. 21.1).

Changes in the volume of production affect standards of living. Monitoring income and expenditure flows provides a check on economic growth, while comparisons between countries is an indicator of competitiveness. Governments use the data to help guide their economic policies. However, collection and interpretation of the figures is not a simple task. Although in theory the three measures should give the same result, defining and aggregating so many items inevitably leads to some discrepancies.

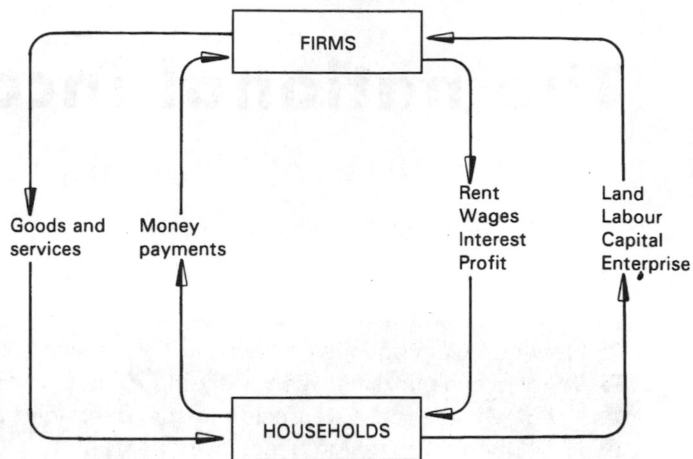

Figure 21.1 Income, output and expenditure flows.

The income account, Table 21.1, shows the total of wages and salaries (by far the biggest item); the profits earned by self-employed people and companies (which may be partly retained for use later by the business and partly paid to the owners); the 'profit' (called surplus) earned by public sector enterprises, whether operated by government departments or local authorities, plus nationalized industries; and finally the income earned from letting property. Included in this is an estimate of the rental value of owner-occupied property as if occupiers paid themselves rent. This avoids the anomaly that would otherwise occur: rented accommodation creates income but owner-occupied property does not, giving a false impression that buying houses instead of renting them makes the nation less well off. The final item, 'imputed charge for consumption of non-trading capital', is a similar adjustment for the use made by assets owned and used by the public sector. Together the figures add up to the **Gross Domestic Product** (income based). GDP figures represent earnings – income from providing goods and services. They do not include payments which are **transfers**, where money is simply redistributed, e.g. benefits or pensions, for which there is no corresponding output.

The biggest category in the expenditure measure (Table 21.2) is consumption – household spending on everything from dustbins to diamonds. Rents for accommodation are included, but the purchase of a new house is classed as investment regardless of whether the buyer is a landlord or owner-occupier. (The price of an existing house, like other secondhand sales, does not involve new production or additional income – it is simply an exchange of assets between buyer and seller. Only the payments made to the estate agency, lawyers, etc., for their services in effecting the purchase count, since these are additional production.)

Table 21.1 Gross Domestic Product (income based)

1991	£m
Income from employment	329 808
Income from self-employment	57 507
Gross trading profits of companies	60 674
Gross trading surplus of public sector enterprises	3 138
Rent	44 092
Imputed charge for consumption of non-trading capital	4 490
Total domestic income	499 809
less stock appreciation	–2 825
Statistical discrepancy	17
Gross Domestic Product at factor cost	**497 001**

Source: *Annual Abstract of Statistics*.

Table 21.2 Gross Domestic Product (expenditure based)

1991	£m
Consumers' expenditure	367 853
General government final consumption	121 899
Gross domestic fixed capital formation	95 442
Value of physical increase in stocks and work in progress	–5 303
Total domestic expenditure	579 891
Exports minus imports	–5 290
Statistical discrepancy	–445
Gross domestic product at market price	574 156
less taxes minus subsidies	–77 155
Gross Domestic Product at factor cost	**497 001**

Source: *Annual Abstract of Statistics*, CSO, 1992.

Investment in new housing and other fixed assets, including buildings of all descriptions as well as infrastructure such as new roads, all comes under the heading 'gross domestic fixed capital formation.' Increases in 'stocks and work in progress' are also investment. Even though they have not yet been bought they have been produced and wages paid to the people who made them, so they must be included in expenditure also. 'Government spending' covers all public sector spending on goods and services such as policing, health care, etc. Welfare benefits are not counted because they are simply transfers. Finally spending by overseas buyers on UK exports is added on and the value of imported goods deducted.

GDP (expenditure based) can be calculated either at **market prices**, which is what the consumers pay inclusive of VAT and other taxes or subsidies, or these can be deducted to give GDP at **factor cost**, i.e. the cost of production (Table 21.3).

Table 21.3
Gross Domestic Product by output

1991	**£m**	**%**
Agriculture, forestry, fishing	8772	1.7
Energy, water supply	28 273	5.4
Manufacturing	104 283	19.9
Construction	33 684	6.4
Distribution, hotels, catering	73 024	13.9
Transport & communications	34755	6.6
Banking, finance, insurance	88 179	16.8
Ownership of dwellings	34 839	6.6
Public administration, defence	34786	6.6
Education, health	49 643	9.5
Other services	33915	6.5
Total	524 155	100
less adjustment for financial services	−27 171	
Statistical discrepancy	17	
Gross Domestic Product	**497 001**	

Source: *Annual Abstract of Statistics*, CSO, 1992.

The output table (Table 21.3) gives a breakdown of GDP by sector. Contributions are calculated on a value-added basis. Since a lot of output is sold on to other firms to be used in their production, a simple aggregate of every firm's output would exaggerate the total. For instance, if a plumber fits a new tap the householder pays for the tap and the work, but the tap has already been recorded as output by the manufacturer. The value added is the plumber's labour and profit margin. Similarly the manufacturer's price included the cost of the iron used to produce the tap, and the iron foundry's price included the cost of coal to smelt the iron. At each stage only the value added is counted as output.

GDP measures the total earnings/output that takes place within the domestic economy, but many businesses also engage in production outside their own country. When Tarmac builds roads in Nigeria wages paid to the local labour are part of Nigeria's national income, but any profits belong to the company's shareholders at home. (NB: This is not the same as exporting goods made at home which have already been included in the accounts.) In the same way Nissan car factories in the UK not only generate earnings for their UK workers but also profits which go back to Japan. To find out UK earnings, GDP must be adjusted by adding on the 'net property income from abroad.' The result is the **Gross National Product (GNP)** (Table 21.4).

To convert GNP to **Net National Product (NNP)** a sum is deducted for 'capital consumption'. This is an allowance for what, in a business, would be called depreciation. If a builder fails to set aside some revenue to

Table 21.4 Gross domestic product and national income

1991	£m
Gross domestic product at factor cost	497 001
Net property income from abroad	328
Gross National Product	497 329
less capital consumption	–63 968
Net National Product at factor cost: **'National income'**	**433 361**

Source: *Annual Abstract of Statistics,* CSO, 1992.

replace plant, the firm's capacity to earn income will decline as its assets wear out. Income is defined as a flow of revenue which can be spent without reducing our wealth (stock of assets). For the nation as a whole this includes social assets, such as roads and housing, as well as capital goods owned by businesses. Deducting the value of output needed to maintain existing assets leaves **national income** or NNP.

INTERPRETING THE STATISTICS

National income is the best measure of a country's spending power, but GDP or GNP are often quoted because estimates of capital consumption differ. Countries vary in size, so for valid comparisons national income is calculated per capita. If the figures are being used to measure economic growth over a period of time changes in the value of money (inflation) have to be considered. The figures can be recalculated in terms of the value of the currency in one particular year and expressed at **constant prices** of, say, 1985 instead of the **current prices** of each year. Variations in exchange rates pose similar difficulties in international comparisons.

Higher income does not always mean a better standard of living. More production generates more income, but not all production increases people's comfort and welfare. Some output may actually diminish welfare. Running a car certainly contributes to our standard of living, but it also produces pollution which has been linked to substantial rises in asthma and related complaints. The output of vehicles adds to GDP but there is no subtraction for the disease caused. Indeed treating the victims generates further income for doctors and drug companies, raising GDP yet again.

Different countries have different needs. Norwegians spend much more on housing per capita than Nigerians, but not all of this represents greater comfort. Norway's colder climate means they have to spend more just to survive. Distribution of incomes also affects welfare. If spending on housebuilding in the UK increased, can we conclude the nation is better housed?

Not necessarily. If the new houses are mostly luxury properties for the richest 5% of the population, while the poorest 25% live in property which is deteriorating, it would be perverse to describe the situation as one of better living standards overall.

For purposes of comparison between countries clearly more information is needed than national income alone, but the statistics do provide a broad indicator of long-term trends. Between 1960 and 1990 the UK slipped from 10th place to 17th in the ranking of OECD countries by national income. Japan over the same period rose from 19th to 5th in the league. Such data suggests a lacklustre performance by the UK.

Economic progress is rarely steady and GDP figures are also used to monitor short-term fluctuations in the economy. The main problem is that it takes time for information to be collected and published, for the government to frame policies, and then for the policies to take effect. The process has been likened to driving a car with a blacked out windscreen and only the rear view mirror to steer by!

DETERMINING THE LEVEL OF NATIONAL INCOME

One of the main short-term aims of economic policy in the post-war era has been to achieve a level of output which keeps resources, in particular the workforce, fully employed. Unemployment lowers living standards and creates social problems but until the 1930s it was not seen as an issue for government intervention.

Economists until then had regarded the economy as a self-regulating system in which market forces acted to maintain equilibrium. If unemployment rose wages would fall, encouraging employers to hire more workers. If workers stayed unemployed for any length of time it must be because they were unwilling to accept jobs at the current wage rate – they were voluntarily unemployed. In the inter-war period mass unemployment undermined confidence in the ability of market forces to keep the economy in balance. Falling wage rates failed to stem the factory closures or the lengthening dole queues. In 1936 a book on the *General Theory of Employment, Interest and Money* by John Maynard Keynes offered an explanation.

The neoclassical economists had tended to focus on supply. The process of producing goods generated the income needed to buy them. Keynes turned the picture upside down. Production, he suggested, responds to demand; if people were not buying goods businesses would close and jobs be lost. Lower wages in these circumstances would only make the situation worse.

To see how this works we will return to the simple model of the economy illustrated in Fig. 21.1. In a system restricted to the two essentials, production and consumption, goods are purchased from the producers by households. Expenditure is equal in value to the goods received. It becomes the firms' revenue, and passed on to workers, shareholders, suppliers etc., providing another round of incomes equal to the level of expenditure. But if consumers choose not to spend all their income, the circular flow will cease to be stable.

Suppose households decide to save some of their income. Expenditure will fall short of current output. Stocks accumulate until firms reduce output to match the reduced demand. They now buy less labour, materials, etc. and so incomes fall. The economy moves into recession. Fortunately savings need not be permanently withdrawn from the flow of funds. They may be recycled as loans to firms who want to invest, e.g. in the purchase of new machinery. Consumer spending is thus augmented by investment spending. So long as total expenditure remains sufficient to buy the total volume of output, incomes can be maintained at their former level.

Savings and investment are not the only withdrawals and injections into the circular flow. When the government undertakes to build roads, or provide public services, this also generates income which injects more spending power into the system. Of course public spending has to be paid for, either by borrowing or by taxes. Taxes reduce spending power and are therefore a leakage or withdrawal from the circular flow. Income also leaks out of the domestic economy when it is spent on imported goods. However, exports have the opposite effect – they bring an injection of earnings from outside. When we include government activity and overseas trade in this way the model becomes more realistic (Fig. 21.2).

Equilibrium depends on a balance between withdrawals and injections. If these are equal, planned expenditure or aggregate demand (AD) will equal total output and be sufficient to maintain production and income levels.

Decisions on saving, spending and investment depend upon thousands of firms and individuals, each acting for their own reasons. There is no guarantee that aggregate demand will match output. If people decide to save more, or firms invest less, withdrawals will exceed injections and expenditure will fall below income. The economy suffers from too little demand and firms respond by cutting output. Less production means lower earnings. Lower income leads to less saving, less spending and therefore fewer imports, as well as smaller tax payments. In short withdrawals shrink until once again they equal the volume of injections and equilibrium is restored – but at a lower level of income.

When AD grows faster than output, through more injections or less withdrawals, firms are stimulated to take on more labour, raise output and so generate more income. Higher incomes tend to increase the volume of withdrawals through savings etc. When equilibrium is re-established, it is at a higher level of aggregate demand and supply. However, if the economy is already working at full capacity, it will be difficult to increase physical output. Without extra resources, more demand will only create shortages, which lead to higher prices. The economy is suffering from demand-pull inflation. In Fig. 21.3 at q^1 the economy has spare capacity. An increase in aggregate demand from AD^1 to AD^2 is matched by increased output, q^1 to q^2. At this point we are close to full employment. The aggregate supply curve, AS, becomes steeper as it gets more difficult to go on increasing production. With continued growth in demand, AD^2 to AD^3, there is little extra output, but a sharp rise in prices, p^2 to p^3.

There are clearly implications for economic policy. By manipulating the level of AD the government can influence the level of economic activity. Demand, as we saw from Table 21.2, comes from consumers, investment, the government itself and exports net of imports. This is abbreviated to:

$$AD = C + I + G + (X - M)$$

Government can raise or lower AD to some extent by changing public sector spending programmes, but it can also influence C and I through taxation and subsidies.

This was the basis of John Maynard Keynes' seminal work in the 1930s. During the inter-war years the country experienced unprecedented

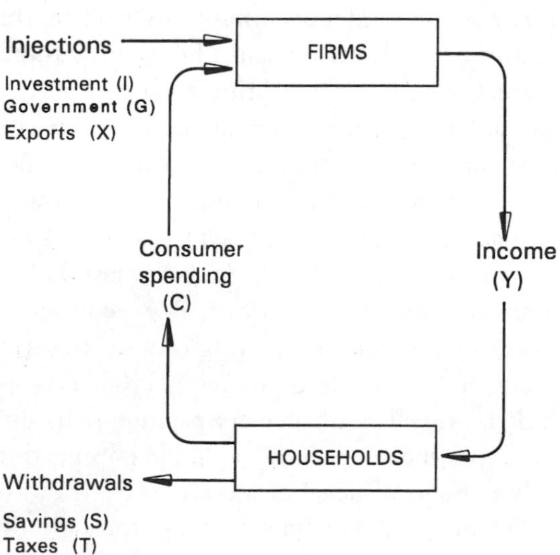

Figure 21.2 Injections and withdrawals from circular flow.

unemployment. The government's income from taxes fell and orthodox opinion said it must therefore cut its spending to balance the budget. Keynes urged the opposite course: let the government spend more, even if it did mean piling up debt. Public spending programmes created jobs and put money into people's pockets. As they spent the money, the demand for goods increased. Rising production meant more jobs would be created and more incomes. With rising incomes the government would receive more in taxes and so be able to cover its debt.

Keynes' argument coincided with the approach of war. Public expenditure rose and unemployment receded. In the post-war era Keynesian thinking became the new orthodoxy. Governments sought to establish an equilibrium level of demand that would ensure full employment without creating inflationary pressures. It was not an easy target to achieve.

Suppose it was estimated that AD fell short of full employment output by £100 m. Public sector spending would not have to rise by £100 m to fill the gap because each £1 spent by the government would enable the recipient to spend more thereby creating more income and raising demand again, an effect known as the **multiplier**. Of course, not all of the extra income will be passed on (if it were, incomes would rise to infinity!). Some will be saved, taxed or be spent outside the system on imports. The size of the

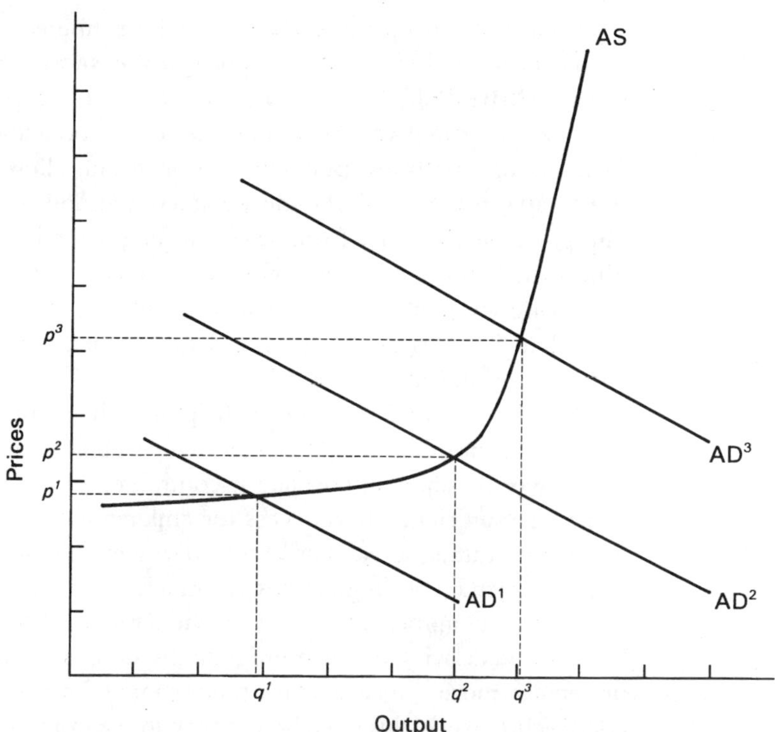

Figure 21.3 Effects of changes in AD on output and prices.

multiplier depends upon the proportion of any extra income which gets recycled through further spending, or the **marginal propensity to consume (MPC)**. The multiplier (k) can be expressed as 1/(1 – MPC). If three-quarters of a rise in income is spent on domestic consumption, k = 1/(1 – 0.75) = 1/0.25 = 4. Each £1 injected will ultimately lead to £4 of additional income. An injection of £25 m would be sufficient to raise equilibrium income by £100 m.

DEMAND MANAGEMENT POLICIES

In the 1950s and 1960s Keynesian views dominated economic thinking and governments assumed responsibility for managing the level of demand. The budget acquired new significance as an instrument of economic policy. **Fiscal policy**, the way the public purse was managed, ceased to be a matter of prudent housekeeping. It was the route to full employment and social justice via alterations in the balance between withdrawals and injections. A deficit budget – more spending than taxes – would boost demand. If the economy 'overheated' a budget surplus, more taxes than spending, could take out some of the excess demand. Post-war governments accepted a wider role and found a new economic theory in Keynes to help them fill that role.

Trying to keep up full employment by running a budget deficit meant the government had to increase the **public sector borrowing requirement (PSBR).** If demand rose too fast and the government wished to slow things down it would operate the reverse policies and either cut public spending or raise taxes. Budgeting for a surplus allowed for **public sector debt repayment (PSDR)** and heralded a period of restraint. Controls on hire purchase and other forms of credit augmented fiscal policy and helped 'fine tune' the economy by acting directly on consumer demand. Achieving growth without inflation or balance of payments problems proved to be a tricky business and governments made frequent switches between throttle and brakes.

These stop–go policies were not helpful to the construction industry which bore the brunt of spending changes. Much public expenditure is contractual and cannot be adjusted; the state is committed to pay pensions and expected to provide sufficient school places for children etc. Cutting current expenditure means cutting services or benefit payments, a politically unpopular move. Capital expenditure programmes are much easier to adjust. Reduced investment has little immediate impact on the level of services. Cutting equipment budgets and delaying building programmes do not mean sacking staff, another unpopular move. Since a large proportion of capital expenditure involves construction work, this puts the industry in the front line.

The construction industry was seen as particularly suited for the role of an economic regulator. It is labour intensive rather than capital intensive, which meant changes in demand quickly fed through into jobs and incomes. There was no need for a slow build up of factories and equipment before it was able to expand. Similarly cutting output meant shedding labour, much of which is casual, rather than decommissioning expensive plant. It is geographically widespread, so the effects are not confined to one area. A lot of the labour is unskilled or semi-skilled, which again makes for greater flexibility in hiring. As an added bonus, in times of expansion construction work could be used to improve the country's infrastructure which would help to stimulate other industries by improving the efficiency of transport systems etc.

From the industry's viewpoint, stop–go policies were damaging. They increased the fluctuations in demand in an industry already subject to cyclical booms and slumps, thus adding to the uncertainty of workloads. Uncertainty discourages investment and hampers planning for the future. In an industry where contracts are often lengthy and may involve a considerable period before construction actually starts, this is particularly difficult. Construction work cannot be stockpiled during slack periods, nor can it be transported to other parts of the country to even out flows of work. When contracts dry up, workers are laid off and the industry declines. For smaller firms, who may only handle one or two contracts at a time, this can mean the business collapses.

More generally fluctuations discourage investment. Whereas labour can be laid off when it is not required, plant lies idle and depreciates. A constantly varying workload discourages investment in training, and makes the industry less attractive to motivated workers who are looking for security and career prospects. It encourages fragmentation and subcontracting as these are ways of spreading the risks. In order to improve efficiency through investment, through training, through the use of industrial techniques and management strategies which extend beyond crisis control, the industry needed a more stable market which would give it the confidence to plan ahead.

Demand management policies seemed to work well in the 1960s, keeping unemployment at bay, albeit with some increase in prices. As people came to expect inflation, they also expected higher wages to compensate. Higher wages meant higher prices and industry became less competitive. Exports suffered and so did jobs. In the 1970s a new problem, stagflation, emerged – a combination of rising prices and rising unemployment. Too much government spending and too much borrowing to finance it was now perceived as dangerous. Demand management policies fell out of favour. The Conservative government elected in 1979 opted for a reduction in public sector activity. They saw market forces as a more effective way of allocating resources. The result was not so much to stabilize the demand for construction work, as to shift the focus from public sector contracts, subject to fluctuations on policy grounds, to private sector clients whose demand varies with the state of the economy.

Rising inflation and a policy of high interest rates in the late 1980s led to a collapse of the housing boom, which had brought sustained growth to the building industry for most of that decade. The housing collapse heralded a more widespread recession and the fall in demand for construction work spread to other sectors. By 1992 a quarter of a million jobs had gone from the construction and building materials industries. Business failures soared, the majority coming from property-related businesses, a grim reminder of the volatility of this sector.

The Labour party has traditionally advocated greater intervention and a more planned use of resources. In the past it has favoured public ownership to achieve this, including setting up a nationalized building industry alongside the private sector, but in the 1980s the shift in opinion away from economic planning in favour of market forces was largely accepted. Programmes for nationalization of building work and other industries dropped out of sight but the party remains committed to a more interventionist approach than the Conservatives. The construction industry could benefit from a government which placed a higher priority on improvements to the infrastructure and social amenities, although any attempt at greater regulation would be resented by many in the industry who take a pride in their independence.

THE ACCELERATOR

The cyclical pattern of activity in the construction industry cannot be explained purely in terms of the stop–go policies described above. Shifts in demand tend to affect capital goods industries, including construction, more severely than the rest of the economy. The impact of changes in consumer demand on industries supplying capital goods is demonstrated in the **accelerator theory**.

The effect is modelled in the figures in Table 21.5, showing how a firm's investment spending varies as its expected level of sales changes. The example makes two assumptions: (a) there is a capital/output ratio of 4, so to produce £10 m worth of goods requires capital assets to the value of £40 m; (b) the capital goods have a lifetime of ten years. To maintain the existing stock of capital therefore requires replacement investment of 10% (i.e. £4 m) p.a.

When consumer demand rises by 20% the additional capital (including buildings) is needed. This boosts investment in capital goods from 4 to 12 units, an increase of 300%. To maintain this level of demand on the capital goods industry, consumer demand must continue to rise. If it stabilizes the firm will not need to increase its stocks of capital goods any further and the investment spending falls back to the level needed to replace worn out capital. Any reduction in consumer demand and the firm has excess capacity, so not even the usual spending on replacement of capital is needed. Surplus equipment can be scrapped, surplus buildings sold. Demand for the suppliers of capital goods dwindles.

This simplified model overstates the effects of the accelerator. In fact firms usually have some spare capacity to meet modest changes in demand. Moreover, any shift in sales would have to be regarded as long term, not just a temporary fluctuation, before investment to raise capacity was contemplated. The extent of the effect would also vary according to the capital/output ratio and the rate at which depreciation takes place. Such modifications dampen the accelerator effect but the model serves to show why capital goods industries are likely to suffer more pronounced swings than the economy as a whole.

TRADE CYCLES

The interaction between the accelerator and the multiplier gives some insight into the phenomenon of trade cycles. All industrialized economies experience a periodic pattern of rapid economic growth followed by a slowdown, stagnation or even recession, until growth picks up again and another boom starts. A number of different cycles have been detected superimposed upon each other. An underlying long-wave rise and fall is named the Kondratiev cycle after the Nobel prize winning Russian economist. This appears to have a wavelength of 50–60 years, with a downswing in the 1870s and again in the 1920s and 1930s. An American economist,

Table 21.5
Demand for capital goods: the accelerator effect

Year	Forecast sales £m	Existing capital stock	Capital stock required	Replacement capital	New capital	Total investment
1	10	40	40	4	—	4
2	12	40	48	4	8	12
3	14	48	56	4	8	12
4	14	56	56	4	—	4
5	12	56	48	(4)	—	(4)
6	12	52	48	(4)	—	(4)

Kuznets, detected another cycle of approximately 20 years duration in construction activity, a fluctuation which he found to be repeated in other countries.

Most noticeable are trade cycles. In the nineteenth century these seemed to be roughly 8–10 years in length. In the post-war period the fluctuations moderated. Some economists believed that demand management policies, based on the Keynesian model described above, had solved the problem but in recent years the pattern of boom and bust has reappeared. The modern cycle seems to move faster, lasting 4–5 years on average.

Many explanations have been offered. Some place the blame on external factors, such as wars, changes in weather patterns, or population movements. Others emphasize the inherent instability of economic systems which makes it unlikely they will maintain a steady path. With multiplier and accelerator effects acting upon each other in the Keynesian model it is easy to see how a small imbalance can turn into a cyclical swing.

If we assume an initial injection, due perhaps to the discovery of new resources such as the North Sea oil fields which sparked off considerable investment spending in the 1970s, this spending will create additional income and, via the multiplier, lead to a period of growth. Rising incomes lead to increased consumption and this brings about a further surge in investment via the accelerator. Multiplier and accelerator reinforce each other creating a boom, but output cannot be expanded indefinitely. As the economy reaches full capacity the growth in demand cannot be met. Imports may be sucked in and domestic prices start rising. Inflation reduces the competitiveness of our goods on world markets. Confidence falters as prices rise, growth in income slows and aggregate demand is no longer expanding. Investment slows, the multiplier magnifies the fall in income and lower consumer demand is picked up by the accelerator to further depress investment; the economy is heading for recession.

The construction industry seems particularly vulnerable to cyclical movements. As a capital goods industry it is liable to the accelerator effect. Industrialists will postpone building plans in hard times. House purchase is also seen as an investment and buyers will delay plans to move if they lack confidence in the security of their jobs and incomes. The time required to build adds to the pattern of shortage followed by over-supply. Construction output mirrors the Gross Domestic Product in its ups and downs, but the rises and falls tend to be sharper (Fig. 21.4). The cycle can turn very swiftly from boom to bust. In the mid 1970s, early 1980s and again in the early 1990s hundred of firms and thousands of jobs disappeared in the industry, some to return in better times, others to be lost permanently. Some of the problems caused by this uneven pattern have been noted above. Bigger companies with large-scale contracts which stretch over two or three years

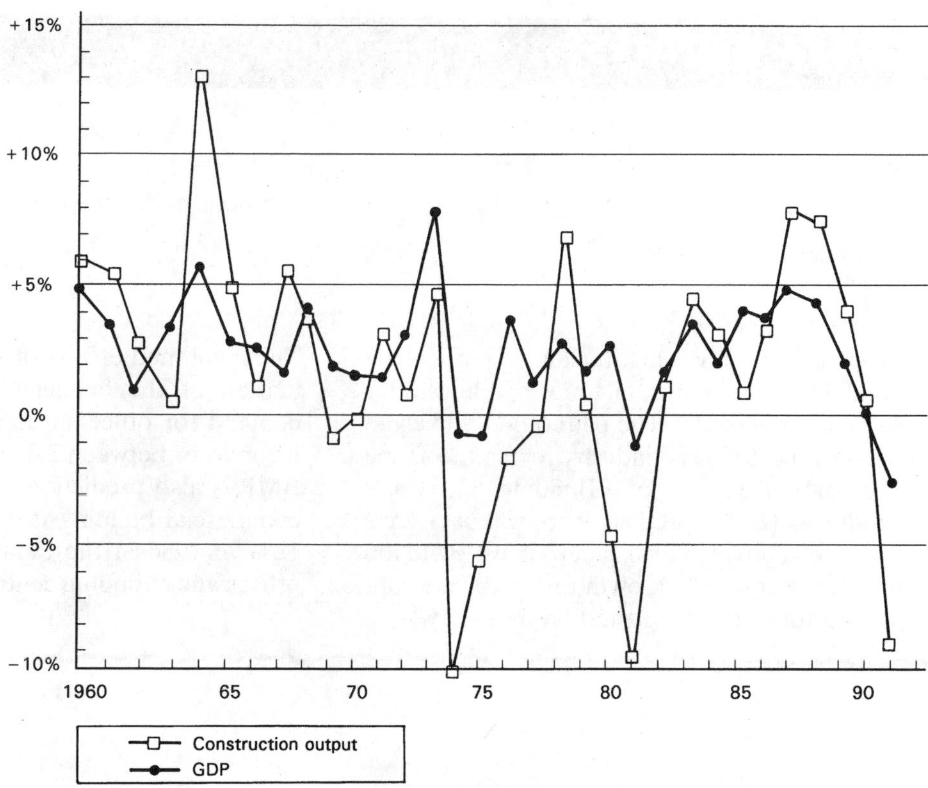

Figure 21.4 Year-on-year change in construction output and GDP.

are insulated to some extent by the work in hand, but even this may not be sufficient in all cases. When larger companies do fail they are likely to leave many unpaid bills amongst their subcontractors which often leads to secondary collapses among the smaller firms.

Keynesian policies of demand management were intended to counteract this cyclical pattern. Critics have suggested that the difficulty of timing the policy measures can result in the opposite effect. Injections intended to boost demand during a recession may not take effect until the economy is already starting to recover. By adding to spending as output approaches its peak the government is aggravating the problem rather than curing it.

CASE STUDIES

A.

1987:
Upswing in the private sector

Over the past few years, building – particularly building offices, shops, factories and houses for sale – has boomed ... The pattern looks likely to continue in 1987. The industry forecast from the National Council of Building Materials Producers (BMP) predicts there will be a 4.5% increase in private housing output in 1987 to follow this year's 9.5% growth to 175,000 new private sector starts, the highest level since 1973.

The combined effects of the Big Bang [deregulation of the financial markets], increasing demand for office space and a growth in real incomes of between 3.5 and 4% mean that the BMP is also predicting a 12% growth in new commercial building in 1987. This follows the 12% in 1986 in the industry's output of new offices and shopping centres.

1988:
Construction machinery unit sales up by over a quarter

Sales of construction and earth-moving machinery have risen by more than a quarter in unit terms this year, probably the biggest increase in demand ever registered ...

The main factor in the steep increase in machinery sales is the construction boom, which is at an eight-year high. Other factors include the replacement of ageing equipment by new machinery and the rapid expansion of the fleet hire sector ...

Sales of rough-terrain lift trucks, for example, rose from 2,830 in 1987 to an expected 4,250 this year.

Building increase 'levelling out'

Construction output will increase by 3% next year before plateauing in 1990 according to the latest report of the Joint Forecasting Committee for the Construction Industries ...

[They] expect the number of private housing starts this year will be 205,000, which represents a peak since 1972. However, private starts next year are forecast to fall ...

Orders for commercial construction have remained high during 1988 but forecasters believe that in 1989 they will not exceed their present levels.

1989:

Building fall hits industrial property

Figures published yesterday by the Department of the Environment showed that the number of homes started by British house builders in September fell by a third compared with the corresponding month last year.

A separate survey conducted by Associated Building Industries shows that many private commercial and industrial developers are postponing starting work on new developments because of the uncertain economic outlook.

Mr Philip Davis, ABI's managing director, said: 'We normally expect to see an increase at this time of year in tender documents sent out by developers so that building work can start before the end of the financial year. This does not seem to be happening.' The number of contracts worth at least £10 m, let or under negotiation, fell by 17% between July and September compared with the same period last year.

Economics director of stockbroker Hoare Govett said the construction industry 'was a good lead indicator of developments in the economy. These figures indicate that a more widespread recession in economic activity is in sight.'

1990:

...a grim roller-coaster ride

Mr Andrew Melrose, construction analyst at SG Warburg, says: 'I do not think investors are aware of how bad conditions have got across a wide area of UK construction. We are forecasting that construction output will fall between 5 and 5.5% this year.'

The fall in house sales in recent months has been accompanied by an almost equally savage fall in sales of commercial property, particularly in London and the south-east. Commercial property values have fallen by 15 to 20% in the City of London ...

SG Warburg forecasts private commercial construction output will fall by 26% next year. The high pace of development has left too many buildings chasing too few tenants at a time when companies face high interest rates and are cutting back on investment.

The fall in commercial property sales also means fewer private sector schemes are being started and contractors have spoken of margins being shredded on private commercial work in the south.

One curtain walling company said it had lost eight orders it expected to win during the two months to the end of August because private sector schemes had been deferred. This represented about 30% of the work it had expected to win during that period.

B.

Economic development theory recognizes the use of leading sectors as engines for growth. In our lifetimes we have seen examples of this theory in the automotive sector with its corollary of highway construction.

The two main conditions for such a leading sector to stimulate growth and create a significant multiplier effect on other sectors of the economy are a large potential demand and a capacity to mobilize domestic resources to meet that demand.

The construction and housing industries are well suited to this role in developing economies. They are the only major sector to rely on basic materials which do not have to be imported. They also use unskilled labour without requiring costly training programmes. They can thus transform considerable numbers of otherwise unproductive persons into productive wage earners and hence consumers.

Based on an article by E. Penlosa in Finance and Development, *IMF and World Bank, vol. 13, no. 1.*

C.

The annual *Spon's Architects' and Builders' Price Book* publishes a guide to price trends in the industry. Additional data can be found in *Spon's Construction Cost and Price Indices Handbook*, both published by E. and F.N. Spon.

Tender prices are the prices clients are asked to pay for work. They reflect the builder's costs and his eagerness to win the job. They are indicative of the state of the market. Cost indices indicate changes in the costs of materials, labour, plant and overheads, including taxes. They are a guide to the cost of carrying out construction work, irrespective of the state of the market.

Reproduced from B.A. Tysoe and M.C. Fleming, Spon's Construction Cost and Price Indices Handbook; *published by E. & F.N. Spon, London, 1991.*

Year	Cost index	Tender index
1970	100	100
1971	109	115
1972	119	145
1973	140	199
1974	166	237
1975	205	242
1976	241	241
1977	275	258
1978	299	299
1979	342	371
1980	410	458
1981	458	467
1982	506	453
1983	537	469
1984	569	495
1985	600	514
1986	631	540
1987	665	608
1988	706	728
1989	759	851

D.

Labour shortage hits recovery

After the worst recession the industry has experienced in decades, figures published by the Department of the Environment showed construction orders for the three months to February 1994 to be nearly a third higher than in the same period a year before. Despite this, only 17% of firms surveyed by the Building Employers' Confederation were operating at close to capacity and growth in output was not expected to exceed 2%.

Sir Brian Hill, president of the Confederation, said 'it is hard to see how the industry could reabsorb all the 500 000 workers lost since 1989'. He pointed out that new technology, improved productivity and the slow growth of output meant the industry was not going to need the same number of workers.

Even though recovery has proved painfully slow, bottlenecks have already developed in some parts of the labour market. Shortages of workers skilled in carpentry, bricklaying and plastering have begun to emerge, putting pressure on wages. Bricklayers' hourly rates have risen from £6 to £7.50. At the bottom of the recession some bricklayers were reported to be getting as little as £4.50, but the rise in wage rates has not brought workers back into the industry fast enough. Worker who went abroad for jobs, or who left the industry to earn money as cab drivers or delivery drivers, are not prepared to give up regular work for the uncertainties of the building industry, and their places have not been filled by new entrants. 'No one's been training properly' complained one manager. His firm, one of the major brickwork specialists in the country, has had to consider turning away contracts because they could not find the skilled tradesmen they needed to do the work.

WORKSHOP

1 Explain why the construction industry is prone to cyclical patterns of demand.

2 Would you expect any variation in the extent to which the following sections of the construction industry suffer fluctuations in demand: civil engineering; repair and maintenance work; building materials; new housing?

3*What are the main problems faced by firms in the industry over the short term as a result of fluctuating demand?

4*Does the cyclical nature of the demand for building work have any adverse implications for the efficiency of the industry in the longer term? Indicate ways in which the industry has adapted to this pattern of production.

5 Why is the construction and housing industry seen as particularly beneficial to the economies of developing nations (see case study **B**)?
Explain the references to 'engines of growth' and a 'significant multiplier effect' in your answer.

DISCUSSION QUESTIONS

- ❑ To what extent have government policies aggravated fluctuating demand for building work? What could be done to create a more stable economic environment for the construction industry?
- ❑ Comment on the data in case study **C** above.

22 The money supply

PREVIEW

- What is money?
- If everyone were paid twice as much money as they currently earn, how much better off would we all be?
- How does bank lending alter the money supply?
- How do interest rates influence the money supply?
- How does monetary policy affect the construction industry?

THE VALUE OF MONEY

In the 1970s unprecedented inflation came to be seen as the most urgent problem facing the government. Efforts to influence the growth of demand, in order to maintain full employment, gave way to policies aimed at slowing the rise in prices. Inflation reached 24% at its peak, which meant in just over three years the value of money would be halved. Concern about prices brought a renewed interest in the quantity theory of money, led by the American economist, Milton Friedman. In essence the quantity theory states that, other things being equal, putting more money into the economy will inflate prices. It offered the prospect of a simple, if painful, solution to the problem of rising prices.

Because money circulates, any attempt to measure the volume of money must recognize there are two variables, the supply of money and the frequency with which it changes hands. Money is exchanged whenever goods (or services) are purchased. The relationship between money, prices and the volume of goods can be expressed as MV = PT, where:

M = money supply
V = velocity of circulation
P = the average price level
T = number of transactions (volume of trade).

This is no more than a truism: the money exchanged for goods and services must equal the value of the items purchased. If the equation is reformulated as:

$$P = \frac{MV}{T}$$

the policy implications become clearer. To restrict the rise in P we must prevent growth on the other side of the equation, which can be done by controlling MV. Assuming that V is constant, i.e. that people's spending patterns are reasonably stable, the target becomes the money supply.

WHAT IS MONEY?

To control the money supply we must know what money is and how to measure it. Money is a means of paying for purchases. A startling variety of things have served this purpose in different societies at different times – precious metals, pieces of paper, bird of paradise feathers, cattle, chocolate, cigarettes and bars of soap. All they have in common is that they were acceptable as payment for other goods. Money is a means of exchange.

In our society that means notes and coins are clearly money, but they are only the small change in our system. Most payments are made by bank transfers, using cheques or credit cards or direct debits. Cheques and cards do not themselves constitute payments, but the transfer of funds from one account to another does, so ordinary bank accounts must be added to the notes and coin.

Many people hold additional sums in savings accounts or building society accounts. Should these be included? Most people would regard them as part of their total holdings of 'money' even though they may not be able to write cheques on their savings accounts nor are able to withdraw the cash immediately to make payments. Can the definition be extended still further to cover premium bonds and savings certificates, or even shareholdings?

There is no obvious place to draw the line. Instead we have a range of assets of varying degrees of **liquidity**. Cash is 100% liquid because it can be used in exchange without delay or risk of loss. Savings accounts are slightly less liquid – we must give notice of withdrawal before we can make use of the money. Shares are less liquid still. They must be sold before their value can be realized; there is a risk of some loss, or even of finding them unsaleable. Shares are not sufficiently liquid to be thought of as money.

Because there is no hard and fast line separating money from other assets a number of different measures are used to monitor growth in the money supply. The narrowest of these is M0, notes and coins, which gives the quickest indication of trends. Today, with many building society accounts offering the same chequebook facilities as the traditional bank accounts,

M2 is perhaps the measure closest to most people's idea of money. It includes personal bank and building society accounts with full chequebook facilities and smaller savings accounts with less than four weeks' notice required. M4 is a broader measure; it includes deposits with more than a month's notice of withdrawal required. (M1 and M3 are no longer used – definitions of money have to change as the way we use it changes!) Clearly bank money, rather than cash, must be the focus of any attempt to control the money supply. Today (1993) M0 and M4 are monitored although there is a suggestion that a new composite measure might be introduced to give a more accurate picture.

CREDIT CREATION

Bank money is created by the banking system itself. Banks originated as places of safe-keeping for gold deposited by customers. As funds accumulated, banks developed into lending institutions. Soon they began to issue their own IOUs, or banknotes. The banks were starting to create credit and so expand the money supply. To see how this works we will model what happens when the banks lend the funds deposited with them.

Initially the bank's balance sheet shows the cash deposited as an asset, balanced by the liability of the same sum owed to its depositors:

	Assets	**Liabilities**
Cash	£1000	£1000 (owed to public)

Holding cash is less profitable than lending it, but the bank cannot lend all the money lest it find itself unable to meet its customers' demand for withdrawals. It therefore keeps 10% of total deposits in cash, leaving 90% available for lending.

	Assets	**Liabilities**
Cash retained	£100	
Loans	£900	
	£1000	£1000

Borrowers use the money to finance spending, e.g. a builder uses a bank loan to buy scaffolding. The scaffolding suppliers pay their takings into the bank.

	Assets	**Liabilities**
Cash retained	£100	
New cash deposits	£900	£900 (new)
Existing loans	£900	£1000 (existing)
	£1900	£1900

The new deposits increase the bank's cash assets and, at the same time, its liabilities. Once again the bank has more cash than it needs to maintain its liquidity. Once again it can lend the surplus, less 10% to cover its increased liabilities.

	Assets	**Liabilities**
Cash retained	£190	
Loans [new + old]	£1710	
	£1900	£1900

The £810 additional loans are spent by the borrower and, when the money is paid into the bank again, will become the basis for the next round of credit creation. The bank is expanding the money supply through its loans. (NB: This is not the same as creating more wealth; whether or not we end up better off depends on whether the money is used to produce more goods!)

The ability of the banking system to create additional credit (money) depends on three things. First there is the amount of funds deposited; the banks can only lend if depositors provide them with the funds to do so. Then there is the amount of cash and/or liquid assets which they need to keep their customers' confidence. Failure to meet demands for withdrawals would destroy the bank's reputation so it must maintain a sufficient ratio of cash to deposits. If the bank increases its cash ratio it will have a smaller proportion of funds available to lend. The third constraint on lending is the demand for borrowing. Opportunities for lending depend on the number of potential borrowers considered creditworthy.

The range of assets held by British banks (Table 22.1) in practice extends beyond cash and loans. By holding a variety of assets banks try to satisfy different requirements of their business. Like any business they seek to make a profit; the highest rate of return is from the loans and advances, but these are also the least liquid. Holding cash (including their balances with the Bank of England) gives liquidity but does not earn any income. If they keep cash to a minimum they will need a second line of defence, other assets which can be converted quickly into cash if necessary. Loans made to discount houses and other financial institutions – often on a very short-term basis, as short as one day – also count as highly liquid. As well as liquidity and profitability, banks require safety, after all the money they are lending is their customers' money. Any loan involves some risk but risks are minimal if the borrower is part of the public sector. Treasury bills, which are short-term loans to the government, also improve the banks' liquidity, while longer-term gilt edged bonds give a secure, predictable income. Traditionally British banks do not invest to any great extent in company shares.

Table 22.1 UK banks' sterling assets, 2nd quarter 1992

	£m
Notes and coin	3 050
Balances with Bank of England	1 609
Loans to discount houses	8 106
Other money market loans	134 547
Treasury bills	2 975
Other bills	10 451
British government stocks	5 416
Other investments	27 429
Advances:	
UK	368 673
overseas	14 078
Miscellaneous assets	32 262
Total sterling assets	**608 596**

Source: compiled from information in the *Bank of England Quarterly Bulletin*, February 1993.

THE NEED FOR MONETARY CONTROLS

Until recently banks and building societies were quite distinct and operated in different spheres. During the 1980s these distinctions became blurred as building societies began to issue cheque books and allow customers to withdraw money without notice. At the same time banks were moving into the mortgage market and competing with the building societies to lend funds to homebuyers. Increased competition and relaxation of the restrictions on the building societies made borrowing much easier. Mortgage queues became a thing of the past and the ready availability of credit helped to fuel the housing boom.

Easy mortgages for house purchase sustained a growth in demand which sent house prices soaring in the late 1980s. In the same way excessive growth in the money supply generally can bring about rising inflation. Monetarists emphasized that inflation is essentially a monetary phenomenon. Demand cannot be made effective unless people have the money to pay the prices asked. (House prices rose more slowly when mortgages were less easily come by – people had to save and wait longer.) Similarly costs, such as wages, will not rise if producers cannot obtain extra money to pay them. Thus monetarists see control of the money supply, or credit, as the primary weapon against inflation.

INSTRUMENTS OF MONETARY POLICY

Controlling the money supply means limiting banks' ability to lend, either by restricting the amount of liquid assets available or by reducing the demand for borrowing. Regulation of the banking system is undertaken on behalf of the government by the Bank of England. This is the central bank, banker to the commercial banks and also to the government. Monetary policy is ultimately decided by the government, although the Bank of England gives advice. Over the years many different policy instruments have been tried. Those mentioned below are the options most often used in recent years.

The simplest method might seem to be to instruct banks to lend less, but the monetarist approach is generally founded on a preference for market forces and dislike of direct intervention. Directives, or instructions, from the Bank of England to restrict lending are rarely used. Deregulation has removed the fixed liquidity ratio, which could be altered to restrict banks' lending capacity. Today banks are expected only to maintain a prudent level of reserves. Special deposits, by which the Bank of England could 'freeze' some of the liquid assets held by the commercial banks, have not been used since the 1970s. Instead efforts have been concentrated on methods that operate via the markets.

Open market operations involve the Bank of England in buying or selling government debts to the public via the financial markets. Selling bonds will mean payments flowing from the purchasers' bank accounts to the Bank of England. The commercial banks, suffering an outflow of funds, will have to reduce their loans to maintain liquidity. In order to expand the money supply the Bank could enter the markets as a buyer. The payments made to the public, once deposited in the commercial banks, provide the basis for an expansion of credit creation.

Government spending, in particular how it is financed, has a major influence on the money supply. Since the mid 1970s increasing use has been made of open market sales to place government bonds with the public. By borrowing in this way, rather than through the issue of short-term Treasury bills (or the even more negotiable IOUs called bank notes!) the government is engaged in **funding** the national debt. This means the debt is raised by instruments which are less liquid, less easily transformed into cash at a moment's notice. Public spending which is financed by selling bonds to the public does not increase the money supply. It just means spending power is transferred from the private sector to the government. The same is true of public spending financed by taxation.

But if the government borrows from the banking system, by in effect running an overdraft with the Bank of England, the money supply will increase. As the Bank honours the payments made by the government it pumps extra cash into the economy. Recipients of the money deposit it in their banks which are then able to lend more and further expand credit. The clear implication of this chain of events is that any government wishing to operate a restrictive monetary policy must not allow its own borrowing to get out of control. A tight monetary policy is likely to mean cuts in public sector spending as well as efforts to fund any **public sector borrowing requirement** (PSBR) from bond sales.

Large-scale borrowing will also tend to push up interest rates. The government is competing with private sector borrowers for the public's savings. To persuade savers to buy bonds rather than put their money into something else, an attractive rate of return must be offered. At the same time high interest rates are a discouragement to borrowers, which reinforces the policy. (It is important to understand that it is the level of real interest rates that matters. If inflation is running at 10%, a nominal interest rate of 10% will mean that the repayment is worth just the same as the original loan. The cost of borrowing in real terms is nil. At 15% the borrower pays just 5% in real terms after allowing for inflation.)

As demand for loans falls off, the process of credit creation will be slowed. The impact of high interest rates on demand for loans is likely to be quicker acting than measures to reduce liquidity ratios. Short-term interest rates are influenced by the Bank of England through alterations in the discount rate, formerly known as the Minimum Lending Rate. Effectively this is the rate at which the Bank will 'bale out' the commercial banks and discount houses if they experience a cash shortage. If this is higher than the going market rate, commercial banks will increase their liquidity, i.e. restrict lending, to avoid incurring losses by having to raise additional cash at a loss. Base rates will also tend to rise in order to minimize the gap between the discount rate and the rates charged by the banks. A rise in short-term rates signals an upward movement in other rates. At present interest rates are the main instrument of monetary policy in use.

MONETARY POLICY AND THE CONSTRUCTION INDUSTRY

Changes in interest rates are particularly damaging to the construction industry. Demand in all sectors of the industry is sensitive to the cost of borrowing. A rise in mortgage interest rates can cause a substantial fall in housing demand. In 1989 a period of rapid growth in the private housing sector was halted by rising interest rates. House sales slowed and prices started to fall. As the economy moved into recession falling incomes and rising unemployment aggravated the problem, but the high cost of mortgages was undoubtedly an element in the decline.

Demand in the commercial and industrial sectors is also linked to interest rates since high borrowing costs mean a higher rate of return will have to be obtained from the investment to make it viable. In the two years

1988–90 the value of new orders for construction fell by nearly 20%. Businesses found themselves faced with high interest rates and then, as interest rates started to fall, lower revenues because of recession. They were unable to afford either to expand or to upgrade their premises. Demand for commercial premises did not react as quickly as the housing market, but the fall was equally severe.

Investment in new buildings is a long-term venture and expectations about future levels of interest rates are just as important as short-term changes. If interest rates are expected to remain high for some time, companies are likely to delay the sort of expansion programmes which would entail new orders for the construction industry. Even the public sector is affected, as many capital projects are financed by borrowing and may no longer be affordable within the budget set. If work is financed out of income rather than by borrowing, as with a lot of minor maintenance jobs, there is still a possible knock-on effect as clients delay less urgent work because they have less cash available.

Construction firms are also borrowers, whether it is just an overdraft to provide working capital for a small business, or a large company using loan finance to build up a land bank. Interest rates affect their costs. When rates go up construction firms find themselves squeezed between falling demand and higher costs. Profit margins are eroded as builders put in lower tender prices to obtain work. Cash flow suffers as clients delay payments and interest charges mount. For many firms loss of liquidity is the prelude to collapse.

MONETARY POLICY AND THE ERM

The fight against inflation entered a new phase late in 1990 when the UK entered fully into the European Monetary System (EMS) and joined the exchange rate mechanism (ERM). By tying the value of sterling to the value of the Deutschmark and the other European currencies, the UK hoped to close one route to price rises. Manufacturers could no longer agree to higher wages or other costs in the expectation that inflation would depress the value of the pound and enable their foreign customers to go on buying their goods. With the exchange rate of the pound fixed, manufacturers who allowed prices to rise would find themselves losing sales. In 1992 the pound was withdrawn from the ERM because it proved impossible to sustain the exchange rate, but with the expectation that it would return at some, unspecified, future date.

Since relatively few construction firms are involved in exporting, does the ERM matter to the industry? Indirectly it matters a lot because it affects interest rates, which in turn affect the demand for construction work. Once tied into the ERM UK interest rates will have to stay broadly in line with

European rates, otherwise people will move money out of the UK to take advantage of better rates elsewhere in the European Union. Large-scale selling of the pound would then push its exchange value below the agreed level. Once in the ERM monetary policy is no longer a matter for the national government alone, so the building industry could not lobby for a cut in domestic interest rates to give it a boost.

Adherence to fixed exchange rates, even a single European currency, could prove to be a significant turning point not only for the economy, but particularly for the housing industry. House purchase in the UK has come to be seen as an investment, a good way of making your money grow. The 'housing ladder' was a way of trading up, with the prospect of doubling your stake in just a few years.

These profits depended on inflation. A mortgage barely affordable today will be easily manageable tomorrow if wages and salaries keep rising. And with house prices rising too, the homeowner can realize his gains and move up the ladder, or borrow against the extra value and have more money to spend. Without inflation, the merry-go-round will cease to turn. If we achieve stable prices, if incomes rise by no more than output, houses may once again be bought as places to live in, as they are in France and Germany, instead of commodities to be traded. The housing boom of the 1980s may enter history as another speculative bubble, which finally burst, and the market of the future be more stable and more closely related to housing need.

SUPPLY-SIDE MEASURES

Whereas Keynesian demand management policies were aimed at getting the economy to operate at full employment without excessive inflation, monetary policy is concerned primarily with prices. Output, the real economy of production and jobs, is not directly addressed. Of course the real economy cannot be divorced from the money aspects; rising prices can destroy competitiveness and with it jobs. For a more direct impact on production, monetarists look to supply-side policies, measures aimed at improving productivity and shifting the aggregate supply curve to the right.

Lower labour costs boost industrial production and encourage job creation. Labour markets are encouraged to be more flexible through reductions in trade union powers, especially closed shops and restrictive practices. Better training is important to improving productivity and job opportunities. Reduced social security benefits ease the burden on government spending, and consequent pressures on the PSBR, as well as increasing incentives to work. Abolition of minimum wages encourages more flexible use of labour.

Deregulation of industry is intended to cut red tape and encourage competition, enterprise and initiative. The privatization programme, from council houses through to the nationalized industries, has had the same objectives. Even within the social sector elements of market forces have been introduced into medical care and education. Overall the supply-side policies, intended to complement monetary measures to control inflation, have re-emphasized the values of enterprise in creating wealth. Improvements in productivity have resulted, but must be set against a widening inequality of rich and poor when judging the outcome of these measures.

CASE STUDIES

A.

Thrift, says the Governor of the Bank of England, has gone out of fashion. Yet in reality it has been driven out by the Government's policy of financial liberalization, as successive Chancellors of the Exchequer scrapped one form of control over bank lending after another, without any complaints from the Bank of England.

How ironic to hear Margaret Thatcher declaring recently that earning before spending was an essential Tory value. Her radical administration dismantled many of the regulations which, by curbing the banks' and building societies' ability to offer credit, promoted serious thrift. As a result British household savings collapsed from nearly 14% of disposable incomes in 1980 to nearer 4% in 1988.

British home owners suddenly found in the 1980s that they could borrow large additional sums on the security of their homes and so increase their spending power. By 1989 the debt of the average building society borrower was nearly 3.2 times his earnings, compared with 2.7 times a decade before.

B.

Construction collapse raises spectre of credit crunch

The construction industry's problems are raising fears of more widespread financial failures. A fall in construction output of nearly 9% in 1991 is expected by some observers to be even worse in 1992. With more people out of work the public purse has less money coming in from taxes; at the same time it has to pay out more in benefits. Consumers with less money in their pockets means less overall spending and reduced orders for industry. This is worry enough for the government, but even worse is the scenario of widespread banking failures following on the collapse in the construction and property development sector.

It is estimated that construction borrowing amounts to over £15 bn and the sector as a whole owes about £40 bn, a quarter of it to UK high street banks. Banks, and more particularly building societies, are also heavily involved in the housing market through mortgage lending. As borrowers default on payments, falling property prices make it hard to recover the debt. Bad debts on this scale could prove catastrophic to the financial system. However, expert opinion seems to view the prospect as unlikely. One analyst was quoted as saying building societies were very secure institutions. Even a 20% fall in house prices could be weathered by 95% of the industry, although some smaller building societies might suffer.

WORKSHOP

1 Explain how bank lending can increase the money supply.

2 What is the relationship between money supply and inflation?

3 Explain the possible causal links between the easy availability of mortgages and:

a a rise in house prices;

b a rise in general inflation.

4 Why might a severe slump in the construction and property sector pose a threat to the integrity of the banking system (see case study **B**)?

5*Discuss the impact of a tight monetary policy on the construction industry.

DISCUSSION QUESTIONS

'Money is a symptom, not a cause. Print fewer bank notes and people will use more credit cards. Limit credit cards and they'll devise another way of paying. The real problem is excess demand.'

'If the money isn't there, people can't spend it – it's as simple as that.'

❑ How simple is it? Does the formula P = MV ÷ T mean growth in money must lead to higher prices? Is spending the result of changes in money supply, or do changes in money supply reflect spending patterns?

23 Housing policy

PREVIEW

- Why does the housing market fail to meet some housing needs in our society?
- Can the housing market be made to work more efficiently?
- Should housing be provided by the state, and if so, why not food, clothing or other necessities?
- Do rent controls help to solve the problem of finding affordable accommodation for people with low incomes?
- Is it more efficient for the state to build houses, to subsidize housebuilding, or to subsidize those in need of housing?

In this chapter we turn from policies aimed at the overall performance of the economy, with the emphasis on output, to look at a particular aspect of the economy, housing, with the emphasis on distribution. Public policy, via building regulations, planning controls, taxes, subsidies and publicly owned housing, intimately affects consumers and providers of housing. We need to examine why housing is a matter for public concern and whether policies are effective.

ORIGINS OF PUBLIC POLICY ON HOUSING

State interest in housing originated in public health concerns. The industrial revolution transformed Britain from a rural into an urban society. The new industrial workforce had to be housed within reach of the mills and foundries which provided their employment. Industrialists and landowners became property developers by default. The rapid growth of the towns was neither planned nor controlled by the authorities, but the problems that resulted from tight packed urban populations soon became matters of public concern. Inadequate water supplies and lack of proper refuse or sewage disposal led to rising death rates and frequent epidemics. The government gave local authorities first the rights, and then the duties, to undertake a variety of necessary tasks. From street cleansing and sewage disposal these were gradually extended. In 1919, local authorities were empowered to build houses for rent.

Since that time housing provision has been revolutionized. At the time of the First World War, 80% of households lived in rented accommodation provided by private landlords. Today two-thirds of households are owner occupiers, with less than one in ten in privately rented housing. Public sector housing, which expanded rapidly in the 1950s and 1960s to reach one-third of the total in 1978, has declined to one quarter as policies have shifted (Table 23.1). Public funds were not used only for building houses to let; owner-occupiers received subsidies through tax relief on their mortgages. Improvement grants have provided assistance to both landlords and owner occupiers. Rent controls, together with security of tenure, seriously distorted market forces in the private rented sector. Despite the preponderance of owner occupation today, market forces are significantly affected by public policy towards housing.

Public involvement can be justified because housing is both essential and expensive, putting it beyond the reach of many. It is one of the three basic requirements of life – food, clothing and shelter – but whereas food and clothing are available in small quantities that can be purchased as they are needed, housing comes in larger units and is built to provide a lifetime's use, or more. Renting allows people to buy accommodation as they use it, but the decline in the rented sector has limited this option. Uneven distribution of income means many poorer people are unable to afford an adequate standard of housing. Poor housing imposes costs upon the whole community (see Chapter 20). Good housing confers benefits which we take for granted, and therefore tend to undervalue. In this respect it is a merit good, one which we may wish to encourage through subsidies.

Table 23.1 Changes in housing stock by tenure (GB)

Total	**Owner occupied**		**Private rented**		**Local authority**		**Other tenures* Housing ass'ns****		**Total**
	000	**%**	**000**	**%**	**000**	**%**	**000**	**%**	**000**
1950	4 100	*29.5*	6200	*44.6*	2500	*18.5*	1100*	*7.9*	13 900
1960	6 967	*42.0*	4306	*25.9*	4400	*26.5*	927*	*5.6*	16 600
1970	9 356	*50.0*	3677	*19.6*	5698	*30.4*	n/a		18 731
1975	10 605	*53.4*	3083	*15.5*	6185	*31.1*	n/a		19 873
1980	11 618	*55.5*	2811	*13.4*	6499	*31.1*	n/a		20 930
1985	13 441	*61.6*	1962	*9.0*	5864	*26.9*	5864**	*2.5*	21 803
1990	15 318	*67.1*	1714	*7.5*	5106	*22.4*	5106**	*3.0*	22 815

Source: *Housing and Construction Statistics*, HMSO. Reproduced with the permission of the Controller of Her Majesty's Stationery Office.

GROWTH OF PUBLIC SECTOR HOUSING

By 1980 local authorities were the largest landlords in the country but until the First World War their role had been negligible. In 1919, in the aftermath of the war, the shortage of decent quality, low-cost housing was so serious that the government passed legislation enjoining councils to provide housing according to the needs of the area. At the same time specifications were laid down which made local authority housing the standard bearer for working-class houses. (This role was maintained when the Parker Morris Report (1961) upgraded local authority standards in line with the general rise in living standards. Regrettably the guidelines were dropped in 1979.) Subsidies from central government aided the councils in their housing and slum clearance programmes. By the outbreak of the Second World War, local councils provided more than one in ten of the nation's houses.

The 1939–45 war resulted in an acute housing shortage. The government's first response had been to prevent the shortage being exploited by a return to rent controls, first introduced in the 1914–18 war. After the war local authorities had to rebuild the housing stock. Nearly a million houses were built in 1945–50, 90% of them council houses. In the 1950s, as the crisis eased, the rate of building slowed, but by the 1960s population growth, rising incomes and higher expectations were combining to create a new housing crisis. Much of the property built in the previous century was no longer fit for habitation. Large-scale slum clearance programmes strengthened the case for state intervention.

Government responded with extra subsidies and exhortations to use industrialized methods. This raised local authority output from one third to one half of new housing completions. Output peaked in 1968 and a change of government in 1970 furthered a slowdown in public sector building (Table 23.2). Demand on the construction industry came in waves, dependent on an amalgam of need, ideology, public awareness, political expediency and financial stringency.

Market forces respond to demand not to needs – in other words the ability to pay is paramount. Markets also respond slowly – building takes time, especially if development takes place in bits and pieces. In short the market could not properly meet the needs of the 1960s. The scale of the problem, moreover, suggested planned and coordinated programmes would be preferable to piecemeal developments, which would have to be restricted in scope by the capabilities and funds of developers. The state, with access to wider information about households, incomes, population trends, etc., was better placed to predict needs, to plan and finance building programmes on the scale required.

Table 23.2 Stock of dwellings: estimated annual gains and losses (GB) (figures in 000s)

	Gains		Losses		Net gain	Total stock
	New construction	Other	Slum clearance	Other		
1956–1960*	290.0	10.8	66.9	20.3	213.6	16 215
1961–1965*	331.3	7.5	77.5	27.2	234.6	17 387
1970	350.4	5.2	88.2	24.9	243.1	18 731
1975	313.0	10.2	61.8	18.0	243.4	19 873
1980	234.4	11.5	31.8	8.8	205.3	21 031
1985	196.9	16.6	10.6	11.4	191.4	21 825
1986	205.8	18.3	8.7	9.6	205.8	22 030
1987	216.3	18.2	8.3	9.8	216.3	22 247
1988	231.8	15.5	6.8	11.9	228.7	22 746
1989	211.0	15.5	6.8	9.2	210.5	22 866
1990	189.8	13.3	7.1	9.5	186.5	22 872
1991	177.8	13.1	6.9	10.2	173.8	23 046

*Annual average over five years.
Source: *Housing and Construction Statistics,* HMSO. Reproduced with the permission of the Controller of Her Majesty's Stationery Office.

Local authorities, already experienced in the building and letting of properties, became major housing providers. The substantial withdrawal of the public sector from the housing market later on, in the 1980s, reflects both an easing of the crisis and a change in ideology. Public sector involvement has been redirected to operate via the market, instead of being an alternative source of housing provision.

AIMS OF HOUSING POLICY

General statements of housing policy are rarely contentious. A 1973 document stated everyone should have 'a decent home with a reasonable chance of owning it or renting the sort of home they want.' The problems arise over how to achieve this. Arguments have centred largely on rent controls in the private sector, and on the role of public sector housing. Subsidies to owner-occupiers have received less attention but can be seen as an equally important influence on housing provision.

In the long term housing policy must look to meeting the changing needs of society. Demographic changes, especially the rate of household formation (Table 23.3), affects the pattern of housing needs. This depends on the age structure of the population and social norms – patterns of marriage and divorce, the desire to establish or maintain independent households by the young and old, etc. Transport facilities and industrial growth will affect job opportunities and commuting patterns. Special needs, whether from disability or social circumstances, must be taken into account. Market forces can be effective in attracting resources to points of

Table 23.3 Estimated trends in household formation (England) (figures in 000s)

	Married couple	Single parent	Single person	Other	All
1971	11 032	1033	3025	865	**15 955**
1981	10 967	1404	3904	1050	**17 325**
1991	10 572	1891	5093	1481	**19 036**
2001	10 142	2336	6354	1771	**20 603**

NB: Figures derived from estimates of population growth divided into households on the basis of historical trends. Decisions to form a separate household are subject to many social and economic influences, including the availability of housing.
Source: DoE 1989 projections.

need in some instances, as the growth of sheltered accommodation for the elderly in the 1980s showed. Other needs, e.g. housing for low-income families, may require state assistance.

In the short term policy must be directed towards making the best use of existing houses. New building adds only 1–2% p.a. to the housing stock, so the supply is very inelastic. However, existing property may be adapted to meet changing needs. Thus many Victorian houses, built for large families and their servants, have been divided into flats to accommodate the smaller households of the present day. Reducing the amount of unoccupied or under-occupied space and seeing that older property is kept in a habitable condition helps to meet current needs.

EFFECTS OF RENT CONTROLS

The most contentious element in post-war housing policy – rent controls – has not promoted good use of resources. Originally they were introduced as a crisis measure to prevent landlords exploiting wartime housing shortages to raise rents. Controls later came to be seen as a way of making sure poorer households could find affordable housing. Controlled rents were unable to reflect market conditions. As a crisis measure they were as effective as any other price controls, but they were ineffective and inequitable in **providing** low-cost accommodation (Fig. 23.1).

They are inequitable because they make landlords, who get less than the market price for their housing, subsidize their tenants. Rent controls take no account of the income of either landlord or tenant. Where the landlord is worse off than the tenant the injustice of this redistribution is very apparent. Rent controls are also inefficient because they do not encourage the best use of resources and do nothing to remedy the shortage of accommodation which brought them into being. High prices are a way of signalling demand and encouraging supply. At controlled rents there is no incentive to provide more accommodation.

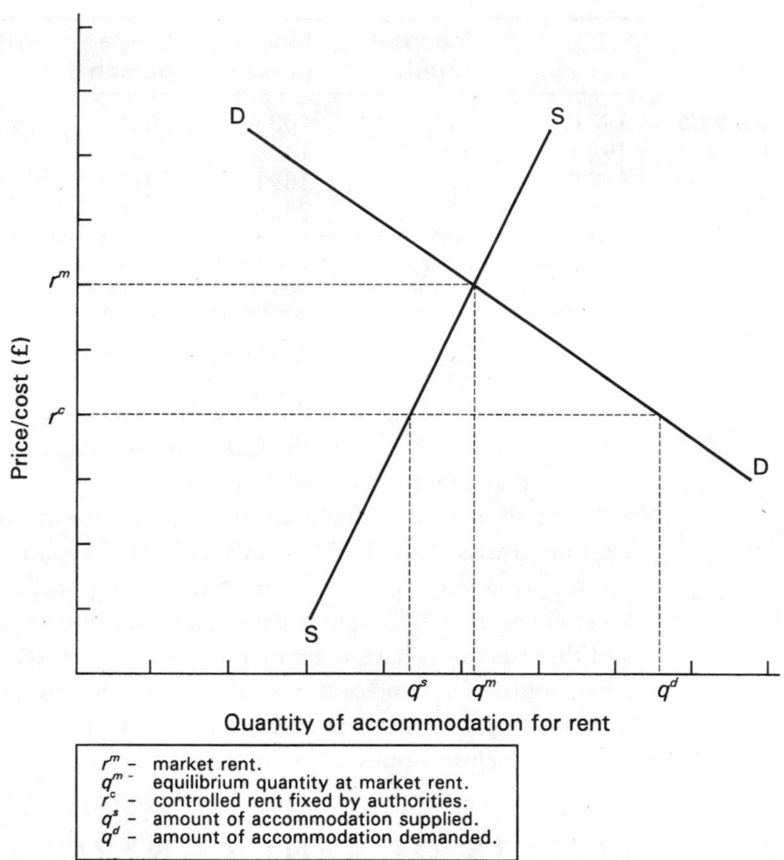

Figure 23.1
Effect of rent controls.

Higher rents make it more profitable to build or convert properties for letting. If rents are kept artificially low the return on capital is reduced and building homes to let becomes unattractive. Existing landlords may find rental incomes are not even enough to cover repair costs and the property gradually decays. Prices of property with sitting tenants are depressed because of the low revenues, so when a tenancy falls vacant the owner may well take the opportunity to sell into the owner-occupied sector where prices reflect scarcity. Thus the longer term effects of rent control are to erode the quality and quantity of housing available for rent. Tenants, who were meant to benefit, may even find themselves evicted to allow the landlord to realize his capital. So, to protect tenants, rent controls have to be accompanied by security of tenure. Letting becomes even less attractive, especially to the owner who envisages an alternative use for the property in the future. The overall result is a further dwindling in the amount of housing to rent.

STATE SUBSIDIES FOR HOUSING

It is beyond the scope of this chapter to trace the intricacies of public sector housing finance, but one or two points can be noted. When local authorities took on the role of housing providers they built and allocated houses on a basis of need. Most councils adopted a points system to assess needs which gave considerable weight to social factors such as overcrowding. Excess demand usually meant potential tenants then joined a queue, or waiting list. Tenants paid rent but this was not intended to provide full funding for the housing programme.

Central government, when it first proposed public provision of housing, promised to share the cost of building programmes. The local authority's liability was limited to the sum raised by a penny rate (local tax) plus the income from rents; government made up the rest. This open-ended commitment by the exchequer led to accusations of council extravagance. Certainly there was no incentive to economize on either design or construction. In 1923 this system was replaced by a fixed subsidy per house. From then on alterations in the subsidies and conditions attached to them were used to influence local authority building programmes. Extra subsidies for slum clearance or for high-rise buildings could guide both the pace and the type of redevelopment.

Subsidies were essential to public housing. Rent controls had been introduced because many wage earners could not afford free-market rents in an era of housing shortage. At controlled rents, building to let became uneconomic. Subsidies were the inevitable next step, if decent housing was to be provided at an affordable price. At first subsidies were expected to be a temporary measure. As the housing shortage was overcome, market prices would fall and neither rent controls nor subsidies would be needed. In practice even subsidized local authority housing went mainly to the better off working class. The problem was that neither subsidies nor rent controls were linked to ability to pay – they were attached to the properties, not the tenants.

The period after the Second World War had followed a similar pattern to the earlier post-war period. First the priority was to reduce the housing shortage, then the emphasis switched to slum clearance. As the shortage eased, general subsidies were withdrawn and authorities urged to charge realistic rents. Decontrol of rents at the top end of the private sector marked the government's retreat from regulation in this area. Higher incomes and changing aspirations meant more and more people sought to become homeowners. By 1965 even the Labour Party, in a document to introduce another boost to public sector building, could describe it as being 'to meet exceptional needs: it is born partly of short-term necessity, partly of the conditions inherent in modern urban life.' The same document looked forward to a period when the 'huge social problem of slumdom and obsolescence' might be resolved and 'the programme of subsidized council housing should decrease.'

Soaring prices and high interest rates in the late 1970s brought the issue of subsidies back into the limelight. An attempt to enforce 'fair rents' and bring the public sector into line with the private sector failed when the Conservatives lost power to Labour. As inflation rapidly overtook rents council housing accounts deteriorated. Tenants' contributions fell from approximately one half to one quarter of costs, while exchequer subsidies rose by the reverse proportions and the deficits mounted. By 1975 the housing debt stood at some £16 000 m.

RECENT TRENDS

The 1980s have seen a radical reappraisal of housing policies. The Conservatives returned to power in 1979, aided by changing economic circumstances and an ideological shift to the right. By 1985 council rents had been raised 128%, an increase 2½ times greater than retail prices; the exchequer subsidy was pruned from over £12 m to under £4 m. Average rents were still under £15 p.w. which the government argued was well below an economic rent of about £28 p.w. As two-thirds of public sector tenants were in receipt of housing benefit, further progress towards economic rents would be largely offset by higher welfare payments. Rising rents caused dissatisfaction among tenants, which was compounded by poor management. Unsatisfactory maintenance and too much bureaucracy showed the deficiency of councils in adapting to a market-oriented role, where customer satisfaction was replacing providing for basic needs as the criterion of success.

This gave added impetus to the right-to-buy policy introduced in 1980. Councils had long had the right to sell, but few used it. Now tenants gained the right to buy and over one million houses were transferred into private ownership in the next decade. This controversial policy was extended to include the sale of whole estates and blocks of flats to private developers for redevelopment, either with sitting tenants or after the compulsory rehousing of tenants. Tenants were empowered by the 1988 Housing Act to opt out of local authority control and transfer their dwellings to a housing association or other approved landlord. The same Act allowed the government to set up Housing Action Trusts to take over and rehabilitate particularly run-down estates before passing them on to approved landlords, housing associations or tenants' cooperatives.

Whether many tenants will opt to remove themselves from council control remains to be seen, but the role of local authorities has been drastically altered. From major providers of mainstream housing, they are increasingly being left with the problems of catering for special needs and homelessness. Reductions in local authority housing stock gives them less flexibility of response. Waiting lists remain lengthy in many areas. While it can be argued that universal subsidies are neither necessary nor desirable today, the public sector has played an important part in setting housing standards and providing a bulwark against real need, roles which are jeopardized by the erosion of its general responsibilities in housing.

Official statistics do not record how many people are without a home of their own. They record only those who have become homeless involuntarily and to whom the local authorities have a duty to provide shelter under the 1977 Housing Act, mainly families with children, the elderly or disabled (Table 23.4). Since 1977 the number of officially homeless persons has trebled. The problem is not confined to the cities, or to any one region. It is a symptom of market failure. People on low incomes are unable to enter the housing market as buyers and alternatives, in the form of rented accommodation in the private or public sectors, have steadily declined.

The costs of homelessness are considerable. Local authorities, with a dwindling stock of council houses at their disposal, have had to find other ways of meeting their obligations. In 1980 1330 households in England were living in bed-and-breakfast accommodation; by 1990 the number had risen eight-fold. The direct costs to the authorities are considerable, but in addition there are indirect costs. The detrimental effects on the people concerned, especially children, create health and social problems. Those who do not qualify for local authority help, many of them young, single adults, often end up living rough. Disease, including tuberculosis which had virtually disappeared in the 1960s, is a growing problem. The homeless become a charge on the community, whether formally via benefit payments or informally via charity or theft.

Table 23.4 Local authority acceptances of homelessness

	England	**Wales**	**Scotland**	**GB**
1978	53 110	3204	6 699	**63 013**
1980	62 920	5446	8 105	**76 471**
1982	74 800	5611	9 303	**89 714**
1984	83 550	4999	9 727	**98 276**
1986	103 560	5965	13 349	**122 874**
1988	117 500	6818	12 601	**136 919**
1990	145 800	8670	15 056	**169 526**

Extracts with permission from John Newton, *All in One Place*, CHAS, 1991.

Home ownership is clearly not an option for those who cannot afford it nor is it the preferred option for many others. People who are single, whose jobs demand mobility or whose incomes are liable to fluctuations may prefer not to have the commitment. In many cases local authority housing is not a particularly good alternative. Recognition of the need for more flexible housing provision has revived interest in private sector rentals. This has meant tackling the twin issues of rent control and security of tenure.

The introduction of shorthold tenures in 1980 gave the landlord the certainty of repossession at the end of the tenancy after 1–5 years. This might have encouraged lets of property which was temporarily vacant, but as rents remained under supervision by rent officers, it did not address the problem of low returns for landlords. The solution proposed in the 1988 Housing Act was to reduce shorthold tenancies to six months and extend the category of assured tenancy (introduced on a limited basis in 1980) to all other new lettings; existing tenancies would continue to run their course under the old regulations. Assured tenants gained security of tenure but rents could be freely negotiated and so reflect market forces.

In areas of housing shortage this meant substantial rises in rents were probable. The intention was to encourage more resources into the rented sector, although high rents could equally well have driven people to consider owner-occupation. Perhaps a greater stimulus to the rented sector will be the disenchantment with home ownership as an investment as the market collapsed at the end of the 1980s. What seems indisputable is that housing remains a problem for those whose incomes are too low for them to compete effectively in the market-place.

SUBSIDIZING OWNER-OCCUPIERS

Owner-occupation has been the most popular form of tenure in the post-war period, greatly encouraged by government policies. Demand is closely linked to income levels and the mortgage market, since borrowing costs are a significant factor in the decision to buy. Government policy on interest rates, although not primarily directed at housing markets, has a major impact upon them. More directly, home ownership has been encouraged through tax concessions. Tax relief on mortgage interest payments was costing in excess of £4000 m by the mid 1980s (Table 23.5). Freedom from capital gains tax is another benefit, one of added value to homeowners at a time when house prices were rocketing. In addition the homeowner enjoys rent-free accommodation, a benefit which had been taxable on schedule A but ceased to be so in 1963. All this added up to a substantial subsidy for housebuyers. Direct comparison between sectors is difficult to calculate, but at the end of the 1970s mortgage relief was costing the exchequer roughly

Table 23.5
Costs to exchequer of mortgage tax relief and subsidies to local authority housing accounts

	Mortgage tax relief			Local authority housing subsidies		
	Average per household	**Current prices £m**	**1985 prices £m**	**Average per dwelling**	**Current prices £m**	**1985 prices £m**
1980	250	1450	2005	246	1274	1762
1985	460	3580	3580	84	381	381
1990	740	6900	5175	155	636	477

Compiled with permission from John Newton, *All in One Place,* CHAS, 1991.

the same as subsidies to tenants. The incoming Conservative government was keen to promote home ownership and by the late 1980s mortgage payers were receiving nearly five times as much subsidy per household as the public sector. The soaring cost of tax relief can be attributed to rising house prices coupled with increasing numbers of homebuyers.

This policy was not without its critics. Mortgage interest relief is not related to needs; the higher the mortgage the more relief can be claimed and since tax relief was initially given at the marginal rate of tax, high-income borrowers gained more than lower-income groups. A limit on the sum eligible for relief, introduced in 1974, reduced subsidies on more expensive properties, but still did not distinguish on the basis of need. Most housebuyers are purchasing within the existing housing stock, so the subsidy did little to direct resources into new building.

The major effect, it has been argued, is to enable buyers to bid up prices, so adding to inflation and putting house prices even further beyond the reach of low-income families. Although ineffective as a subsidy to aid the needy, even harmful in its effects on the market, tax relief is well entrenched. Now that two-thirds of the population are families with an interest in the housing market governments are wary of removing the subsidy. However, as prices rise further, the upper limit of relief, set at £30 000 in 1983, means that it is losing some of its importance and may be allowed to dwindle away. Restriction of relief to the standard rate of tax has also helped to moderate the cost to the exchequer.

'A PROPERTY OWNING DEMOCRACY'

During the 1970s and 1980s house prices rose on average 3.5% faster than the general rate of inflation. First-time buyers were worst affected since they had no property to sell. With the rented sector in decline, the fear of house prices escalating out of reach and the prospect of large capital gains from property ownership meant more and more would-be homebuyers were eager to take out the maximum mortgage they could obtain and get 'a foot on the ladder'.

Price rises could not be sustained at these levels indefinitely. Sooner or later there would have to be a pause while incomes caught up or the first-time buyers who sustained the market would be squeezed out (Table 23.6). This cyclical pattern of rapid price rises followed by a slowdown had been experienced in the 1970s and again in the 1980s. The increase in home ownership meant there were more people than ever before with mortgages, many taken out at the height of the boom. When recession lowered incomes many found themselves unable to keep up payments. Repossessions soared and for some the dream of home ownership turned into a nightmare of homelessness.

Housing policy and the effect on house prices is clearly significant to the construction industry, but it is equally important to the wider economy. In the 1980s escalating house prices affected the labour market, helping to fuel wage demands. Regional differences in house prices have reduced the mobility of labour, as have shortages of rented accommodation in the public and private sectors. The rising asset value of houses and the favourable tax treatment given to owner-occupiers has encouraged the use of savings for home purchase, for home improvements and to trade upwards on the housing market. This, it has been argued, has diverted much needed funds away from investment in industry. Borrowing against the rising value of their homes also enabled owners to free capital for consumption, thus adding to inflationary pressures.

As we entered the 1990s with the market plunging into recession new problems emerged. For the first time in half a century property owners have experienced falling house prices, falling not just in real terms but in nominal terms. High interest rates, coupled with rising unemployment, saw many purchasers unable to keep up their mortgage payments. With falling prices many buyers could not even recover their debt by selling their homes. Forced sales of repossessed property flooded the market and

Table 23.6 Affordability of mortgages for first time buyers (UK)

	Ratio of mortgage advance to recorded income	Net mortgage outgoings as % of income net of tax
1970	1.96	18.6
1975	1.94	22.8
1980	1.67	25.5
1985	2.00	29.4
1990	2.27	40.5

Reproduced with permission from John Newton, *All in One Place*, CHAS, 1991.

depressed prices further. With so many houses already on the market, new developments suffered a sharp decline and many builders were forced into bankruptcy. The knock-on effect of the housing recession was widespread. All sorts of goods associated with moving into a new home – electrical goods, furnishings, DIY supplies, etc. – suffered too. Even homeowners whose mortgages were paid off felt poorer and less secure as a result of the fall in property values, and cut back their spending accordingly. The housing market came to be seen as the key to consumer confidence, a pivot on which the whole economy turned.

Such effects are difficult to quantify or prove beyond doubt, but it is clear that an unregulated market, which lurches from boom to bust, is not in the best interests of either homeowners or tenants. It is equally clear that housing policy has not yet succeeded in providing everyone with a 'decent home' and a reasonable option between purchase or rental. Too often policies have been forced into an ideological strait-jacket. Housing is a basic commodity which shapes our lives for years to come. Policies dominated by the political needs of short-term governments do not necessarily serve the best interests of the population.

CASE STUDIES

A.

Inner city blues turn to bliss

Since 1990 council estates across Britain have felt the wind of change. Hornsey Lane, in Islington, north London, is one of 20 estates now run by a management board drawn from the tenants. The new managers are on the spot; when tenants have problems they don't have to wait for a bureaucratic council system to respond, they go and bang on a manager's door.

Hornsey Lane was a blackspot, crime-ridden, decaying, a no-go area where milkmen and paper shops refused to deliver. Built half a century ago, its fabric was run down, its social disintegration symbolized by a series of killings which gave the area notoriety. Today it is clean and cared for, a desirable estate where people talk to their neighbours. It is gaining a new sort of fame as a model of regeneration. Improvements in the built environment have gone hand in hand with a renewal of the social environment.

A determined campaign led by Mrs McCarton, a tenant of over 20 years' standing, persuaded the council to undertake a multi-million pound renovation scheme in the late 1980s. Central heating was installed in place of open fires, kitchens modernized and the most cramped flats demolished. External walkways disappeared, replaced by gardens and landscaping and a new community centre.

Then, in October 1991, the tenants voted to take responsibility for the running of their estate. The eight-strong board received training in management and financial skills. They employ maintenance workers and reckon to get minor repairs dealt with within 24 hours. They deal with disputes between neighbours. They take a pride in their community. The only thing left to the council is to collect the rent.

B.

Wrong priorities in housing

The scheme announced at this week's Conservative Party conference to convert council rents into mortgages is a logical extension of the 'right to buy' policy launched a decade ago ... But it would be unwise to expect the scheme to have more than a marginal impact on Britain's highly inefficient housing market. It will not reduce homelessness nor curb the excessive fiscal privileges of middle-class homeowners. It will do little, if anything, to improve job mobility. It will not revive the still sickly private rented market ...

The government's dogmatic desire to curb the role of local authorities and boost private ownership at any price has militated against rational decisions in housing. It has obscured the fact that the public and private sectors should play a complementary role. Home ownership most clearly makes sense for middle and upper income families in stable circumstances ... In most economies, the private rented market is regarded as the best for people on moderate and variable incomes who need to remain mobile ...

WORKSHOP

1 Comment on the statistics shown in Table 23.1

2 On what grounds can it be argued that housing is a merit good?

3*Using supply and demand diagrams, analyse the effects of rent controls on the market for rented accommodation.

4 Examine the case for and against the use of mortgage interest relief to encourage home ownership.

5 Consider the advantages of renting as opposed to buying accommodation.

6 Discuss the role of local authorities in providing social housing.

DISCUSSION QUESTION

❑ 'Homelessness is an inevitable symptom of market failure; it can only be tackled by government intervention.'

24 International trade

PREVIEW

- What are the gains from international trade?
- Why do we import goods that we can produce at home?
- How important is foreign trade to the construction industry?
- Why has interest in overseas construction markets grown in recent years?
- Are the problems on overseas contracts the same as for domestic contracts?
- How is the European Union affecting the construction industry?

REASONS FOR INTERNATIONAL TRADE

The earliest theory of international trade saw it as a struggle between nations to win the lion's share of the world's wealth. Mercantilists measured wealth by the amount of gold which a nation accumulated. Exporting goods earned gold and made a nation rich. Importing goods made a nation poorer because it then had to pay out its gold. Adam Smith developed a more dynamic view of trade. He saw that wealth was not fixed, but could actually be expanded by trade. Wealth, he argued, is not how much gold you have, but how many goods and services you can enjoy. Prosperity depends on the flow of production, the size of the national income. Specialization increases production and, through trade, can raise living standards.

Viewed in this light trade ceases to be a war, with every transaction having winners and losers. Instead it becomes an exercise in mutual benefit. What makes trade possible is that countries are endowed with different resources. Climate, soils, mineral deposits, populations, skills, knowledge, all these and many other factors contribute to a country's ability to produce a variety of goods. If each nation specializes in the goods which it is best suited to produce, and exchanges its surplus with others, more goods can be made available to all. The argument is an extension of the basic case for the division of labour.

As long as each country has an absolute advantage in some goods, the benefits of specialization and exchange are easy to see. If one country has fertile soils, a mild climate and efficient farmers while another is endowed with poor soils, a harsh climate and innovative technologists it is clear that the former is better suited to agriculture, the latter to industry. Since farm goods and manufactured goods are needed by both it makes sense for them to specialize and trade instead of trying to be self-sufficient.

THEORY OF COMPARATIVE ADVANTAGE

If one country is better able to produce both types of goods than the other it is not so obvious that there is anything to be gained from trade between them. The theory of comparative advantage, which is illustrated in Figs 24.1 and 24.2, shows there is still likely to be mutual benefit.

For a given amount of resources Industria can produce either 20 000 units of farm produce or 16 000 units of manufactures. If resources are split equally between them, the combined output will be 10 000 units of farm

INDUSTRIA: A given quantity of resources can produce **EITHER**
20 000 units of farm goods **OR** 16 000 units of manufactures

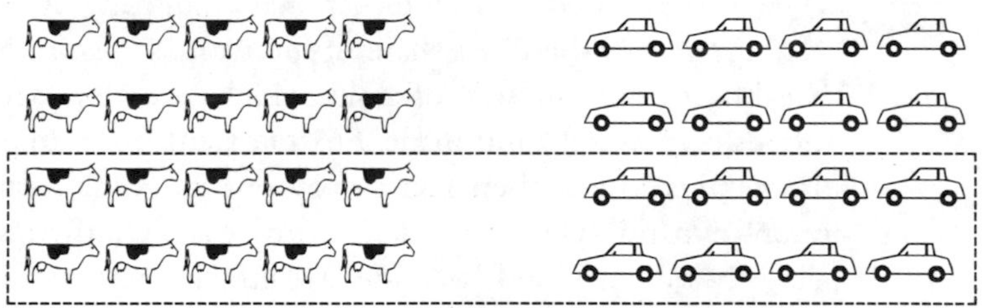

With resources divided equally Industria produces

10 000 units of farm goods **AND** 8000 units of manufactures

RURITANIA: Resources can produce **EITHER**
18 000 units of farm goods **OR** 6000 units of manufactures

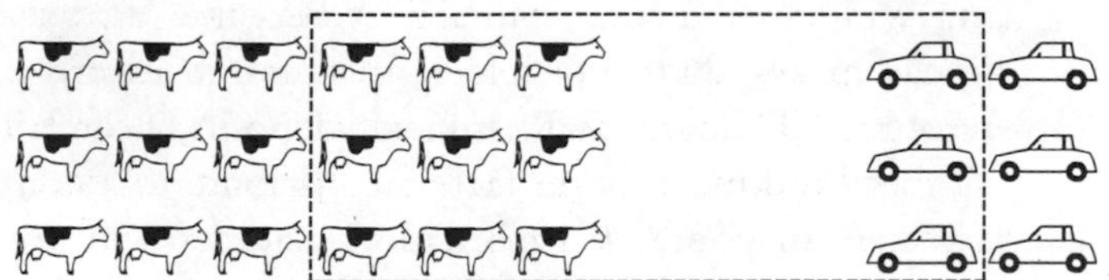

With resources divided equally Ruritania produces

9000 units of farm goods **AND** 3000 units of manufactures

Figure 24.1
International output before specialization.

INDUSTRIA: Comparative advantage in manufactures

With 1/4 resources in farming and 3/4 resources in industry Industria produces
5000 units of farm goods **AND** 12 000 units of manufactures

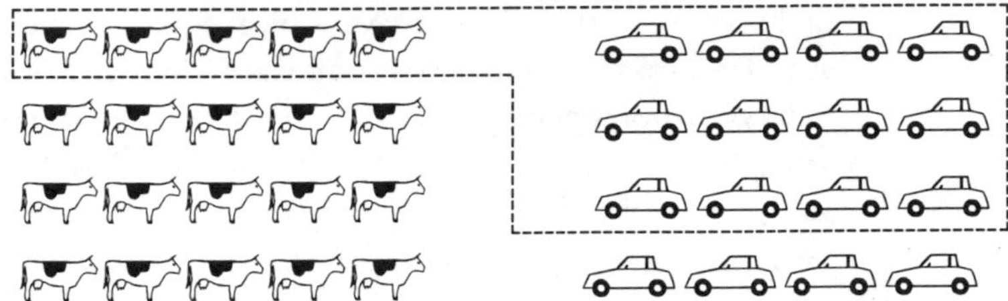

RURITANIA: Comparative advantage in agriculture

With 5/6 resources in farming and 1/6 resources in industry Ruritania produces
15 000 units of farm goods **AND** 1000 units of manufactures

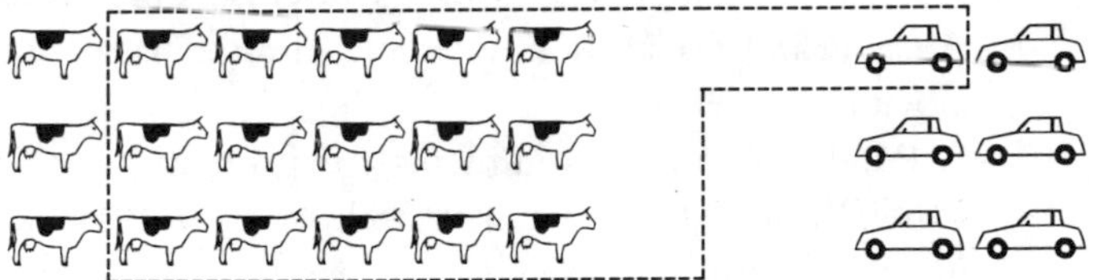

Figure 24.2
International output after specialization.

produce and 8000 units of manufactures. If Industria wants more manufactures they must switch part of their resources and sacrifice some agricultural output. Each unit of manufactures costs them 1.25 units of farm goods.

Ruritania is a less efficient economy. The same amount of resources produce only 18 000 units of farm produce or 6000 units of manufactures, or 9000 units of food plus 3000 units of manufactures. Their exchange ratio is 1 : 3, manufactures to farm goods. Clearly Ruritania could get more manufactured goods if it trades them for farm produce with Industria than if it switches resources at home. Industria also benefits because its manufactures will 'buy' up to 3 units of farm produce in Ruritania instead of the 1.25 units they are worth at home. Provided the exchange is negotiated at a price in between 1.25 and 3 units of farm produce per unit of manufactures, both countries stand to gain. Of course this gain can only be realized if there are extra goods available. The potential gain in overall output can be seen in Fig. 24.2. If Industria devotes three-quarters of its resources to manufacturing and Ruritania concentrates five-sixths of its resources in agriculture, their combined output in both categories is greater than before specialization. This ignores any improvements in productivity that specialization may bring.

The theory shows the potential advantages of specialization. It does not explain why so much trade today consists in exchanging nearly identical goods. The UK, for example, both imports and exports cars with France and Germany. If all markets were perfect and all behaviour totally rational, such inconsistencies might disappear. In the real world of imperfect markets, non-homogeneous goods, partial knowledge and consumer quirks of taste, such trade continues. It gives us a great range of choices. Moreover, in a world of national currencies and political barriers to trade, deals may be struck through an elaborate system of barter, or counter-trade, whereby exporters accept goods which they then have to sell on, to clinch their own sales.

Despite the benefits of trade, restrictions on the free movement of goods will continue as long as nations seek to maintain separate political identities. Governments pursuing the national interest will always be reluctant to allow their countries to become over-dependent on others for what are seen as essential goods or services. The much criticized Common Agricultural Policy is an example; it aims to secure Europe's food supplies, albeit at higher prices.

Purely economic arguments for protection are less easily justified. It is generally accepted that new industries may need a period of protection from established competitors as a temporary measure, to allow the infant industries time to establish themselves. A similar case may be made for geriatric industries to allow for rationalization while the industry is reorganized and surplus capacity redeployed. Most other arguments for protection are less soundly based.

The commonest complaint against foreign goods is that they undercut the home producers and destroy jobs. Provided import prices are not artificially low, perhaps through subsidies, this argument has little justification. Low prices are as good a reason for buying from a foreign source as from a domestic supplier. Buying more cheaply allows us to enjoy more goods; trade barriers would protect jobs for some at the expense of higher prices and lower living standards for all. A low price indicates the foreign producer has a comparative advantage. Where this arises from lower labour costs it is sometimes argued this gives the overseas producer unfair advantage, but using low cost resources is a good reason for specialization. The country with high labour costs can compete through the use of technology and by specializing in production which does not require cheap labour. This may mean some industrial reorganization but in a world of constant innovation change is inevitable.

GROWTH OF OVERSEAS CONSTRUCTION MARKETS

As construction work is generally produced *in situ*, the scope for trade is limited. The industry enjoys a degree of natural protection from overseas competition because of the obvious difficulties of transporting structures. Nonetheless international trade is becoming increasingly important to the industry. The analysis in the previous section is consistent with trade in three categories: firstly natural materials, which are unevenly distributed around the world, a case of absolute advantage; secondly, manufactured components, where specifications and costs may vary between countries; thirdly, and perhaps most importantly, there is expertise, whether technical or managerial.

Most building materials are bulky and heavy. Transport costs are correspondingly high. For this reason traditional buildings have used local materials and developed appropriate styles of construction. The result contributes to the pleasing diversity of regional character displayed in our towns and villages. Yet even in medieval times some needs were met from overseas – such as stone for cathedral building imported from France. Today the use of industrial products, like steel and concrete, together with improvements in transport, mean imported materials and components are widely used. From tropical hardwoods to German curtain walling and Swiss lifts, the range of foreign produced building materials is extensive. Construction imports exceeded exports by over £2 bn in 1988, a reflection of the high level of activity within the industry which domestic materials producers were unable to keep pace with.

International markets for construction expertise have grown enormously in the past two decades. Since management skills are the industry's main resource, this is the most important aspect of overseas trade for contractors. In the early 1970s UK firms had relatively small interests in overseas work. Foreign projects accounted for approximately 5% of contractors' turnover. In the later 1970s the international market exploded. The main opportunities arose in the Middle East, as the under-developed, oil-rich nations put their wealth into improving the infrastructure of their economies. Contracts were issued for building roads, airports, hospitals and other public amenities. This coincided with a downturn in public spending in the UK so many companies were eager to develop alternative markets for their skills. Foreign contracts trebled in value – by 1978 they accounted for over 11% of all new work.

A similar, if less spectacular, expansion took place in other under-developed countries, especially in Africa. Middle East oil revenues played a part in funding this work too as the world's financial institutions sought to lend the funds being deposited with them. Again the emphasis was on developing the infrastructure, on schemes to improve transport and communications and facilities for irrigation and power generation.

By the early 1980s these markets were slowing down. The Middle East had completed many of its major programmes. Poorer countries were finding the world economic situation less favourable. Inflation and high interest rates made existing debts increasingly burdensome and new projects impossible to fund. On the supply side international markets became increasingly competitive. Contractors from the Far East, including Korea, Japan and India, appeared on the scene. Countries which had previously been providers of low-cost labour to work on major contracts were now bidding to organize those contracts. The UK industry was suffering from high rates of domestic inflation and an unfavourable exchange rate which made it less competitive and overseas earnings fell back.

The debt problems of the poorer nations led to a reappraisal of alternative markets in the richer nations of the industrialized world. Opportunities for profit were actively looked for, from housebuilding in the USA to mineral exploration in Australia. Some companies sought to broaden their range of activities, e.g. Costain's and Taylor Woodrow's acquisitions of mining interests in the USA. Others looked to maintain their core activities but on a wider geographical base, e.g. HAT Group's extension of its repair and maintenance services into the American market.

Building up overseas operations has often been done by the acquisition of foreign companies, but mergers are not the only route to foreign contracts. However, knowledge of local markets and management with expertise in local conditions are essential. In the words of one major contractor: 'It is important to get into the information chain early', something best done by 'someone resident in the country ... who is well-known and understands the local problems' (quoted in *The Modern Construction Firm*, Hillebrandt and Cannon.)

Joint ventures are an alternative to taking over a foreign firm. These are set up for specific projects, usually in the form of a limited liability company with a fully integrated management drawn from the partners involved. The loyalties of those involved are thus directed towards the project itself, rather than their individual parent companies. One UK contractor's comment 'there are few problems of different methods of operation, which may mean that contractors' behaviour is becoming more alike' is indicative of the trend towards internationalism of major construction markets.

Large schemes, such as the Docklands redevelopment, can involve many firms. In general such consortia are less closely integrated than a fully joint venture. They do not work for a common profit, but retain individual responsibility for their own profit or losses. In other respects they work in a similar manner, pooling their skills and working cooperatively, not competitively. Consortia of this nature are likely to become more widespread if the numbers of multi-million pound contracts continue to grow. (In the early 1990s, with the world economy in the doldrums, there is a dearth of mega projects.) Major contracts are often beyond the resources of a single company. A consortium can spread the risks and pool resources of several firms without creating a permanent organization.

OPPORTUNITIES IN THE EUROPEAN UNION

Ironically the European market saw less activity by UK firms after our entry into the then EEC in 1973 than before. This was partly a reflection of the lucrative opportunities offered in the Middle East. By the 1980s European companies were more active in British markets than vice versa, although the total volume of such work remained small. The advent of a single European market, officially in place since 1993, made little immediate impact.

Even before the single market links were being forged between British and European companies, ranging from one-off organizations such as Transmanche, the Channel Tunnel grouping, to longer-term links through the acquisition of shareholdings in companies across the Channel. It is a two-way business, which also gives continental firms easier access to British markets. For UK firms the advantages include wider access to finance and to technology not available locally, as well as the hope of reciprocal opportunities to bid for contracts in the partner's home market. The potential rewards are considerable. The EU construction market is estimated to be worth about £300 bn, six times the size of the UK market. With a number of Eastern European nations looking for membership, this market is likely to be even bigger in the future.

Creating a single market has meant harmonizing standards for products, procurement rules and building practices. Over fifty technical committees have debated standards for products ranging from adhesives to ventilation systems. Products which conform to the specifications laid down in the Construction Products Directive gain the 'CE' mark of approval and are then deemed fit for sale throughout the Union. Manufacturers benefit from wider markets and potential economies of scale, while builders gain an extended choice of products. There are opportunities here for improvements, but, as with all new products, there are risks too if builders fail to appreciate their uses and limitations.

The rules for procurement in the public sector aim to ensure open competition on all major contracts by laying down rules for advertising and awarding jobs (see appendix to this chapter). Open tenders are the rule, except in certain, specified circumstances.

Harmonization of building practices includes setting minimum safety standards and establishing common liability rules to cover responsibility for defects. A single market also means free movement of personnel, so the recognition of professional qualifications across borders is another issue. In general a minimum of three years' study on a recognized higher education course will equip the successful student to enter professional practice in any member country, but there are provisions for 'top-up' qualification where necessary, including a period of supervised working. Architects have enjoyed mutual recognition since 1985, but other professions have been less easily harmonized. Quantity surveyors, for instance, hardly exist as such outside the UK; in France the nearest to their role is either *l'économiste de la construction*, normally working on the architect's brief, or *le metteur*, whose job is to measure completed work, a job of lower status than the British quantity surveyor.

Harmonizing the rules is a necessary condition of creating a single market, but it is not sufficient. Before firms can compete effectively they must come to terms with the different styles of working, of operations and contracts, which characterize each region. A single market is not the same as a uniform market. Designers and builders must recognize local preferences. French houses are commonly built with cellars, German office workers do not like sealed windows and central air conditioning systems, and so on.

Overseas contracting can offer big rewards. It also carries a higher level of risk and requires careful market research. Knowledge of local conditions is essential – not only the physical conditions, but the economic, political and cultural conditions also. French contractors use approved lists of material suppliers; German contractors do not. Germany is a federation of states or *Länder* and the majority of contractors are locally based. Subcontractors have the responsibility to design and manage their packages. Performance guarantees are usual and the emphasis is more on quality than cost control. In France smaller projects may be undertaken by trade contractors working on separate contracts. It is a very competitive market with a mixture of large and small firms similar to the UK.

INTERNATIONAL COMPETITION

Increasingly construction expertise is only part of what is required. In less developed countries especially, ability to put together a complete package, including finance, is often crucial to winning contracts. Despite Britain's historic links with so many one-time colonies, the construction industry has performed poorly in world markets. In 1981 there was only one UK contractor in the world's top 30. Like many other industries in the UK there has been lack of foresight and too insular an approach.

The Japanese penetration of British and European markets provides an instructive contrast. Japanese management trainees undertake lengthy periods of study abroad before helping to set up subsidiaries in the country concerned. Clients are often other Japanese firms, but the actual work is subcontracted to local firms. It is essentially management contracting based on a knowledge of local conditions acquired by thorough training and research, using local labour and materials. Concern for local sensibilities in one instance even went so far as to lead to the abandonment of an ambitious scheme to build a complex for Japanese pensioners in Spain. Where UK firms compete for contracts on price and quality Japanese contractors have been accustomed to building long-term relations with their clients (and subcontractors). In such a culture performance is paramount, price is secondary.

Japanese construction shares with Japanese manufacturing a commitment to quality which is reflected in its working practices. Buildings are not notable for their aesthetic qualities but they are built to high specifications. Since Japan is located in one of the world's earthquake zones structural standards are vital. The construction process emphasizes consistency and reliability. Contracts are completed, down to the last detail, on schedule. Meticulous planning, large numbers of closely detailed drawings and standardized methods help to ensure that site workers know what is required of them. Schedules are programmed to allow for catching-up time if needed, so that progress is maintained. Operatives attend daily tool-box meetings, or briefings, to inform them of their targets and any special requirements. They are encouraged to look for ways of improving their working practices. Japanese building is not cheap but offers a high degree of reliability.

In contrast to the high degree of pre-planning which typifies Japanese construction, American methods emphasize speed and price. Productivity is generally high compared to the UK, due more to design and organization rather than differences in actual labour efficiency. A cost study (reported by *Building,* 30 April 1993) of a contract for a combined operations centre at Heathrow revealed some unexpected results, summarized in Fig. 24.3. The project was priced in the UK and the USA. To allow for differences in standard practices between the two countries the US estimates were prepared on three sets of specifications: one identical to the UK, another modified sufficiently to meet US codes, the third modified more substantially to conform with normal US methods but without altering the appearance of the building.

Surprisingly the carbon copy was marginally more expensive in the US, a difference of £70 000 (less than 0.5%) on a price of £17.1 m.

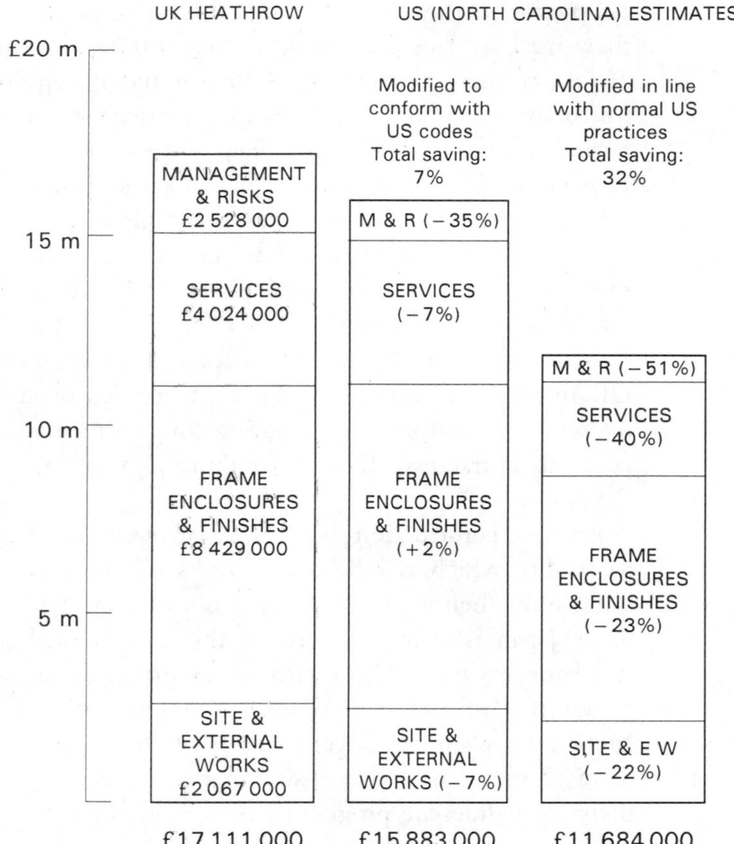

Figure 24.3 Breakdown of UK and US construction costs for airport operations centre.

Conforming to US codes produced some cost savings. Ribbed slabs in place of flat slabs for floors and roof reduced the weight of these elements and allowed further savings on foundations and columns, but other modifications, such as laminated glass in place of toughened glass in the curtain walling, added to costs. Overall the savings amounted to 7%. When the design was fully 'Americanized' costs fell dramatically by 32%. To put it another way, American clients could have half as much building again for their money.

The main sources of savings were the simplified design and layouts, which lightened loadings and reduced foundation work, plus more use of standardized components in place of custom-designed elements. The substantial time savings from this, both on preliminaries and site time, meant that estimates for management and contingencies, costed on a percentage basis, were also greatly reduced.

In the construction world of today, technical expertise is not enough. The industry, as always, must meet the demand if it is to succeed, but demand is constantly changing. Japanese firms, in contrast to their Western counterparts, spend 1% of their turnover directly on product innovation and basic research. The industry must be flexible enough to match the needs of tomorrow, not those of yesterday, if it is to compete in international markets, but the fragmented structure of the UK industry may prove a handicap in developing long-term horizons.

APPENDIX: COMPETITION IN THE SINGLE EUROPEAN MARKET

Public sector construction contracts in the EU have been subject to directives aimed at opening the market to competition on an EU-wide basis since 1971. With the single market officially in place from the start of 1993, the rules have tightened up. All public works contracts over 5 m ECU (approx. £3.4 m) have to be advertised in the *Official Journal* and *Tenders Electronic Daily*. There is a prescribed format for tender notices which specifies the information that must be given, to include tendering procedures, nature and extent of work, financing arrangements, etc., and a timescale. In cases of urgency shorter time limits may be imposed, but these are only permitted in exceptional circumstances.

Tenders may be open or select, but the criteria for disqualifying firms from selection are specified to prevent discrimination. The criteria for awarding the contract are simple: the lowest bid or the 'economically most advantageous tender'. The latter might include such matters as technical merit, completion time or running costs and would have to be justified by the awarding authority. Where firms feel discrimination has occurred, they can lodge a complaint and, if necessary, seek redress either through the national courts or via the Commission.

Based on *Public Procurement and Construction, Official Publication of the European Union.*

CASE STUDIES

A.

Reproduced with permission from Contract Journal, *17 March 1983.*

The director of the Cement Manufacturers Federation (CMF) commented that the proposed quality assurance scheme had been favourably received by the government in the wake of its own White Paper, published in the previous year (1982) on standards, quality and international competitiveness. It was planned to launch the scheme in conjunction with a major promotion to boost sales of British cement.

Cheap foreign cement was threatening to undercut domestic suppliers, but, warned the CMF, quality was not as high as the domestic product. Currently the CMF price for a minimum 15 tonne load delivered in Westminster is £42.50 per tonne. Imports of West German cement arriving in London Docks were selling for approximately £10 per tonne less, but the CMF stressed that their price included technical support services and a timed delivery.

Although relatively little foreign shipments were coming in to London as yet, there was a concern that large amounts might be forthcoming in the future from Eastern Europe and that this could pose a threat to the domestic industry. Any reduction in capacity would not only affect jobs, but would place the construction industry in a position where it had to rely on imports whose availability could not be guaranteed.

The CMF expressed itself unwilling to provide technical assistance to purchasers of foreign cement, which often included additives which were not easy to detect. 'British cement makers will do everything they can to ensure that customers who are loyal to British cement suffer no long-term disadvantage in relation to any of their competitors who are seeking to undermine them by the use of imported cement,' pledged CMF.

B.

When is a brick not a brick?

Reproduced with permission from National Builder, *October 1989, the then monthly journal of the Building Employers Confederation.*

According to Richard Smith, at the Brick Development Association, 'The case of bricks gives a good example of the complications facing the committee [for the harmonization of standards].

'In the UK a brick can still be treated as solid if 25% of its surface is perforated, but in Germany this figure is 15%, so if German standards were adopted British bricks would be disadvantaged.

'There are problems too with test methods. In Britain, for instance, dimensional deviations are measured by lining 24 bricks in a straight line side by side and measuring them with a steel rule. The Germans use callipers and the Dutch use a gauge.

'As the chairman is a German professor it seems likely we will end up using callipers, but what seems ridiculous is that the committee is prepared to propose a tolerance of plus or minus 4% without having an agreed method of testing.'

C.

Carry on working ...

More than a month after the invasion of Kuwait some British companies are continuing to work on contracts in Iraq. In practice it has not proved simple to cut off all commercial contact at the stroke of a legislative pen, as the case of Mivan Overseas Ltd, one of Northern Ireland's leading construction companies, demonstrates.

Official diplomatic advice is that British expatriates could run the risk of being taken as hostages if they down tools. Mivan has joined several British companies in telling its expatriates to carry on working normally as a result ... rules and regulations have effectively frustrated the export of goods and material from the UK to Iraq since August 7, but Mivan is one of several companies that already had its material in Iraq and was thus unaffected by the ban.

The company believes it has discovered a 'grey area' in the regulations regarding payments due for work in progress since the invasion of Kuwait ... [The Bank of England] said it was prepared to approach the issue of payments on a case by case basis.

D.

Table 24.1 provides details of the value of contracts obtained for UK construction work overseas (adjusted to current prices).

ble 24.1
ntracts obtained (£m at current prices)

	1979	**1980**	**1981**	**1982**	**1983**	**1984**	**1985**	**1986**	**1987**	**1988**	**1989**	**1990**
ope	109	120	31	57	56	96	136	100	906	329	188	105
ddle East	652	513	632	789	758	596	250	261	125	153	158	200
a	89	132	276	367	204	254	197	172	191	114	104	353
ica	203	353	354	841	447	341	203	174	158	133	161	320
erica	274	236	434	463	557	746	301	844	1334	1496	1417	1211
eania	58	112	143	263	192	221	183	153	630	91	211	289
tal	**1385**	**1366**	**1870**	**2780**	**2274**	**2456**	**1491**	**1704**	**3345**	**2316**	**2239**	**2478**
% of all v work	13.8	13.5	17.4	24.2	16.8	6.7	9.7	9.9	15.1	8.8	8.2	10.2

rce: *Housing and Construction Statistics,* HMSO. Reproduced with the permission of the Controller of Her Majesty's Stationery Office.

WORKSHOP

1 What are the main arguments justifying protection? In the light of these, comment on case study **A**.

2 Comment on the figures shown in case study **D**.

3 Your company is considering expanding its operations overseas. Prepare a report for the board outlining the main benefits for the firm.

4*What are the main problems likely to be encountered by construction companies operating in overseas markets?

5 What is a joint venture? Why are joint ventures often used by companies working overseas?

DISCUSSION QUESTION

❑ To what extent is there a European construction industry? Discuss the likely impact of the single market on the UK construction industry.

Glossary

Accelerator The tendency for investment spending to increase when the economy is growing at a faster rate than aggregate demand; similarly when the economy slows, the investment falls more rapidly than total expenditure.

Agglomeration The geographical concentration of an industry which benefits individual enterprises within that area.

Capital Resources which have been accumulated as a result of past production, including buildings, plant and tools, transport systems, improvements to land, etc.

Cash flow Comparison of receipts flowing into a business with payments out during the same period; not the same as profit and loss account.

Ceteris paribus Assuming all other variables remain constant.

Competitive tendering A method of introducing competition into contracts to undertake work, by inviting firms to put in a price for the job. Invitations may be open to all-comers or restricted to a pre-selected group.

Complementary goods Goods or services which are in joint demand. A change in the price of one will affect sales of both; for example, a fall in computer prices will extend the demand for computers (move along demand curve) and increase demand for software (shift demand curve to right).

Conditions of demand All the factors which influence the level of demand, i.e. the position of the demand curve.

Consumer sovereignty The power of consumers, in a competitive market, to determine what is produced through their choices of what to purchase.

Contestable market A market where the costs of entry/exit are very low. In the absence of significant entry barriers, the threat of competition will keep costs and profit margins low even if the existing producer is in a monopoly position.

Cost-benefit analysis Appraisal of projects with significant externalities to determine the net benefit to society of the scheme. It involves putting a financial value on intangibles such as noise or leisure.

Cost effectiveness analysis A more limited appraisal to ensure output is obtained for the minimum cost.

Cross-price elasticity The impact of a change in price of one good on the demand for another. It is measured by dividing the percentage change in price of A into the percentage change in quantity demanded of B.

Demand (effective) The desire to buy backed by the ability to pay. The quantity demanded is normally inversely related to price; this can be shown graphically by a demand curve.

Derived demand Demand for an input (factor of production) which is determined by the demand for the final output.

Diminishing marginal returns (or **variable proportions)** A situation where the addition to output from the last worker employed is less than the output added by the last but one employee. This assumes that no change has been made in the amount of capital (fixed factor) as the workforce has been expanded. Diminishing marginal returns can apply to any variable factors used in conjunction with a fixed factor.

Direct costs Costs that are attributable directly to the volume of production, such as site labour and materials. Also known as **variable costs**.

Economic goods Any goods which are scarce, and which therefore involve an opportunity cost.

Economic rent Earnings received by the owner of a factor of production in excess of the amount needed to induce him/her to keep the factor in its current use.

Economies of scale/Diseconomies of scale Reductions in unit costs experienced as a result of large-scale production. Diseconomies occur when unit costs start to rise as output is expanded, usually due to management costs.

Elasticity The responsiveness of quantity demanded/supplied to a change in one of the variables affecting demand/supply. See also **cross-price elasticity, income elasticity** and **own-price elasticity of demand**.

Enterprise Willingness to undertake the risks of organizing and financing production. Sometimes classed as the fourth factor of production.

Equilibrium A situation where consumer demand is exactly matched by suppliers' offers for sale. There is no pressure on prices to move either up or down.

Equity The value of an owner's stake in a business. **Negative equity** occurs when the value of an asset, e.g. a house, falls below the outstanding debts secured against the property.

Externalities Costs, or benefits, that result from production but are not taken into account by the producer or consumers because they fall upon people outside the transaction.

Factor cost The costs of factors used to produce goods.

Factors of production (land, labour, capital) The resources needed for production, whether natural, human or man-made.

Financial economies Tendency for larger enterprises to obtain funds at lower cost through their ability to spread risks and offer greater security to investors.

Fiscal policy Policies on taxation and public sector spending. By intentionally budgeting for a deficit (surplus) governments can manipulate aggregate demand to stimulate (restrain) economic growth.

Fixed capital Capital goods which contribute to production over a period of time.

Fixed costs Costs which are not related to volume of output but remain constant in the short term. Also known as **overheads**.

Free enterprise economy An economy in which individuals are free to organize production in response to consumer demand. Also termed a **market economy**.

Free goods Goods where supply is sufficient to meet demand at zero price, i.e. there is no **scarcity**.

Gearing The extent to which a business relies on borrowing in proportion to the size of its **equity.** Highly geared companies have a high ratio of debt to equity finance.

Gross development value The capital value of a building. It is estimated by multiplying the projected net income by the reciprocal of the target rate of return.

Gross Domestic Product (GDP) The aggregated value of all output produced in the economy during the year.

Gross National Product (GNP) The aggregated value of output produced by British nationals during the year.

Income effect The effect of a change in purchasing power on the quantity demanded of a product.

Income elasticity The extent to which demand changes in response to a change in income; measured by dividing percentage change in income into percentage change in quantity demanded.

Inferior goods Goods whose consumption falls as consumers' incomes rise. They display a negative **income elasticity of demand**.

Infrastructure Assets which contribute to an economy's productive potential by being made available for public use, such as roads, schools, housing, etc. Sometimes called social capital.

Investment An addition to the nation's stock of capital assets.

Labour Human resources used in production.

Labour productivity The output achieved by a given input of labour.

Land All the natural resources available for production.

Liquidity The ease with which an asset can be acquired and disposed of without losing value during the time it is held, or incurring costs for the transaction. The most liquid asset is money.

Managerial economies Reduction in unit costs through greater efficiency of management as a business increases in size.

Managerial utility Setting objectives to satisfy managers' desires, e.g. for status. These aims may not maximize profits.

Marginal cost The addition to total costs when output is increased by one extra unit.

Marginal efficiency of capital (MEC) The rate of return on investment at which one additional unit of capital investment just breaks even.

Marginal physical product The additional goods or service produced as a result of increasing inputs by one unit.

Marginal propensity to consume The proportion of any change in income which is spent on consumption.

Marginal revenue (marginal revenue product) The addition to total revenue from the sale of one extra unit of output.

Market economy An economy where the great majority of decisions about production and consumption are made by individual producers and consumers interacting in the market.

Market failure Where goods are either produced to excess, or too few are produced to maximize social welfare, usually because the market fails to recognize all the costs or benefits involved (see **externalities**).

Marketing economies Reduction in marketing costs per unit as volume of production increases.

Merit goods Goods which are likely to be under-consumed because the benefits extend beyond the consumer, e.g. education.

Mixed economy One which uses a mixture of centralized controls and market forces to allocate resources and distribute goods.

Monopolistic competition A market where suppliers have the monopoly of their product, which is distinguishable from others usually by brand identity, but where there is competition from similar products and an absence of serious entry barriers.

Monopoly A market which is controlled by a single supplier.

Multiplier The ratio of the final change in equilibrium level of **national income** to the initial change in aggregate demand which caused it.

National income (net national product) Total value of the nation's earnings in the course of a year, minus the value of capital assets used up during the year.

Normal goods Goods the consumption of which rises as consumers' incomes increase.

Normal profit The rate of return adequate to maintain production, but no more.

Normative statements Statements which contain opinions. These can be agreed with or disagreed with, but not proven or disproven.

Oligopoly A market in which a small number of suppliers dominate sales, each aware that the others will react to any price changes.

Opportunity cost The best alternative which could have been produced or consumed had the producer/consumer not chosen as they did.

Optimum output Production at a level which results in the lowest unit cost.

Own-price elasticity of demand The impact of a price change on the quantity demanded, measured by dividing the percentage change in price into the percentage change in quantity.

Perfect market A market where prices respond freely to supply and demand, without any individual or organization having the power to influence them.

Positive statements Factual statements which are capable of being shown to be true or false.

Production possibility frontier The limits to production set by the resources available, assuming resources are fully and efficiently used.

Productivity The amount of output achieved for given volume of inputs. See also **labour productivity**.

Public goods Goods whose benefits are shared by everyone, without loss of individual satisfaction, such as defence.

Public Sector Borrowing Requirement (PSBR) The amount by which public sector spending exceeds income from taxes or the sale of goods and services, which sum must be financed by borrowing.

Public Sector Debt Repayment (PSDR) Excess of public sector revenues over expenditure, which allows reduction of debt.

Quasi-rent A payment in excess of the minimum required to retain the services of a factor of production due to temporary shortages.

Rent (a) Income derived from property. (b) Rewards paid to a factor of production in excess of its **transfer earnings**.

Resources See **factors of production**.

Satisficing Management based on achieving adequate performance over a range of targets rather than maximum performance for a single target.

Save To refrain from consumption.

Scarcity When supply is insufficient to permit everyone to consume freely as much as they want.

Short run A period during which firms do not have time to alter their capacity by changes to their **fixed capital**.

Social capital See **infrastructure.**

Substitute goods Goods which are alternatives in the consumer's eyes, competing for the available spending power.

Substitution effect The tendency to prefer less expensive goods in place of more expensive goods, ***ceteris paribus***.

Sunk costs Costs which cannot be recovered, e.g. the monies spent on research or on assets which have no resale value.

Technical economies of scale Reduction in unit costs as a result of changed technology associated with large-scale production.

Transfer earnings The rewards which a factor of production could receive from employment elsewhere; hence the minimum reward needed to retain the services of that factor in its current use.

Transformation curve A graphical representation of the effects on output of transferring resources between two alternative types of production.

Utility The benefit gained from consumption of goods or services.

Variable costs Production costs which vary with the volume of output.

Working capital The stocks of raw materials, components and finished goods that are maintained as part of a business, including uncommitted resources in the form of net cash balance.

Further reading

The following list is not intended to be a full bibliography. It simply offers suggestions for readers who wish to explore the subject further. These range from straight economics texts to books for professional practitioners in the construction world. I hope that this work will have helped to bridge the gap that sometimes seems to exist between them.

The choice of introductory books in the field of general economics is huge. The two titles selected below are both modern and wide-ranging. They will take beginners well beyond the scope of the present volume.

A. Anderton, *Economics*, Causeway Press, 1991.

J. Sloman, *Economics*, 2nd edn, Harvester Wheatsheaf, 1994.

Textbooks which relate economics specifically to the construction industry include:

G. Briscoe, *The Economics of the Construction Industry*, Mitchell Publishing in association with CIOB, London, 1988.

S.D. Lavender, *Economics for Builders and Surveyors*, Longman, 1990.

J. Raftery, *Principles of Building Economics*, BSP Professional Books, Oxford, 1991. (This book is less of an introduction to economic principles, more of an exploration of building production problems from an economic viewpoint.)

M. Warren, *Economics for the Built Environment*, Butterworth Heinemann, 1993.

Books on the construction industry from an economic viewpoint include the two titles below by Hillebrandt and Cannon that present the outcome of an investigation into the objectives, strategies and decision processes of large construction firms. The first draws on theoretical analysis from the disciplines of economics, management, financial control and sociology and tests these against practice by means of in-depth interviews with senior executives and board members. Evidence from the interviews is presented in the second volume.

P.M. Hillebrandt and J. Cannon (eds), *The Management of Construction Firms: Aspects of Theory*, Macmillan, London, 1990.

P.M. Hillebrandt and J. Cannon, *The Modern Construction Firm*. Macmillan, London, 1990

Hillebrandt's earlier work resulted in two books which no subsequent writer on the subject can ignore. Like others I have drawn upon the analysis in these:

P.M. Hillebrandt, *Economic Theory and the Construction Industry*, Macmillan, London, 1974.
P.M. Hillebrandt, *Analysis of the British Construction Industry*, Macmillan, London, 1984.

A historical perspective is given in the following titles. The first deals with the 1970s and 1980s, focusing on changes in the relationships between firms, the labour force, the professions and the client. The second looks at growth and change in the industry in the 1950s and 1960s.

M. Ball, *Rebuilding Construction*, Routledge, London, 1988.
M. Bowley, *The British Building Industry*, Macmillan, London, 1966.

The remaining titles concern a variety of issues related to land use and the construction industry. Apart from the first two they are not 'economics' books, but they all touch on issues relevant to the problems of utilizing resources which lie at the heart of economics.

P.N. Balchin, G. Kieve and G. Bull, *Urban Land Economics and Public Policy*, 4th edn, Macmillan, London, 1988.
J.M. Harvey, *Urban Land Economics*, Macmillan, London.

F. Harris and R. McCaffer, *Construction Plant – Management and Investment Decisions*, Granada Publishing Ltd, 1982.
J.W.E. Masterman, *An Introduction to Building Procurement Systems*, E. & F.N. Spon, London, 1992.
A.W. Millington, *An Introduction to Property Valuation*, Estates Gazette, 1988.
I. Seeley, *Building Economics*, 3rd edn, Macmillan, 1983.

Building economics, or design economics, is not an analysis of economics concepts but an examination of cost-effective design and construction. Many relevant issues, such as the use of capital and labour, industrialized building methods, etc., are explored here. The next title is similarly concerned with technical rather than economic issues, but again is related to the efficient use of resources.

D.J. Ferry and P.S. Brandon, *Cost Planning of Buildings,* 6th edn, BSP Professional Books, Oxford, 1990.

The Chartered Institute of Building produces a series of Occasional Papers which provide students with up-to-date discussions of particular topics. For example:

A.A. Kwayke, *Fast Track Construction*, CIOB Occasional Paper No. 46, 1991.

CIRIA (Construction Industry Research and Information Association) is another source of up-to-date research reports and assessments of the industry. Individual professional bodies disseminate information through conferences and publications. Trade journals often have articles of interest concerning economic aspects of the industry, including market forecasts. Academic research is catered for in the journal *Construction Management and Economics* published by E. & F.N. Spon.

HMSO publications include a large number of official statistics. Those most directly concerned with the industry are to be found in *Housing and Construction Statistics*, an invaluable source of data.

Index

Page references appearing in **bold** refer to case studies.